THE REPRESENTATION PROBLEM FOR FRÉCHET SURFACES

BY

J. W. T. YOUNGS

Fellow of the John Simon Guggenheim Memorial Foundation

INDIANA UNIVERSITY

TABLE OF CONTENTS

PREFACE

The subject matter of this paper deals with the representation problem for a collection of Fréchet varieties known as Fréchet surfaces (see 1.2)[1]. On the other hand, the representation problem may be stated, and a particular form of a solution offered for all Fréchet varieties. This being the case, the discussion will begin on general lines and the need for specialization will become apparent later, when it will be observed that the present frontier of mathematical knowledge and the general form for a desired solution appears to dictate a specialization of the Fréchet varieties to be considered.

It should be stated that an active effort is made to do something more than offer a solution to the representation problem. In the first chapter a measure of attention is directed towards an indication of how the pattern of research on this problem developed during the decades following the initial major assault on the problem by Kerékjártó [5][2]. This does not mean that the comments in question are devoted to tracing the history of the subject as such — for this and further bibliography the reader may consult Youngs [17] — but rather that the emphasis is on a discussion of the methods of attack and the directions along which these methods evolved. In addition, an attempt is made to indicate those difficulties which initiated the introduction of the topological tools here employed. Consequently, it is hoped that the first few paragraphs will serve two ends. For those who may wish to learn about the problem but are uninterested in the details of the solution, the discussion will serve, perhaps, as a means of acquiring a relatively painless understanding of the problem and a fair measure of intuition for the subject. For those who will go on to study the proofs, it is hoped that these paragraphs will provide an overall picture of the situation before one is plunged into details.

It should be mentioned that active use will be made both of analytic and algebraic topology; specifically, of the cyclic element theory and the

---

1. *Numbers in parentheses refer to paragraphs in this paper.*
2. *Numbers in brackets refer to the bibliography.*

Čech and locally compact cohomology theories. A general reference for the former has been Whyburn [10], for the latter, a forthcoming book by Eilenberg and Steenrod (see [3] and, in this connection, Cartan [2] where the locally compact theory is discussed in terms of Lefschetz groups).

The reader is referred to the index of terms and the index of symbols at the end of the paper. These are employed to eliminate the excessive use of cross references.

CHAPTER I

THE PROBLEM, METHODS OF ATTACK, DIFFICULTIES

1. The representation problem for
Fréchet varieties

<u>1.1</u>   The representation problem is easy to state in spite of the fact
that it appears to be somewhat difficult to solve.

The basic notion is that of a *mapping* $f : X \to Y$; that is, a
continuous transformation from one space into another. The notation and
terminology of transformations to be employed here will follow Lefschetz
[6, pp. 2 – 3]; unless otherwise stated, however, the spaces X and Y will
be metric, and for brevity a statement such as *"f is a mapping"* will sometimes be employed without explicitly naming X or Y. In addition to these
common conventions, a mapping $f : X \to Y$ is said to be *trivial* [*non-trivial*] if
$f(X)$ is *degenerate;* that is, a single point [*non-degenerate;* that is, consists
of more than one point]. Moreover, $f : X \rightrightarrows Y$ means that f is from X *onto*
Y; that is, $f(X) = Y$, while $h : X \approx Y$ means that h is a *homeomorphism*
from X *onto* Y. (The notation $X \approx Y$ is read X is homeomorphic to Y).

It should be mentioned that the symbols $\to$ and $\rightrightarrows$ are much too
useful to be reserved exclusively for the purpose indicated above. If X is
a metric space and $x_n \in X$, $n = 0, 1, 2, \cdots$, then the notation $x_n \to x_0$ means
x_n *converges to* x_0; that is, if ρ is the distance function in X, then
$lim \ \rho\{x_n, \ x_0\} = 0$. If $f_n : X \to Y$ is a mapping, where X and Y are
metric, $n = 0, 1, 2, \cdots$, then the notation $f_n \to f_0$ means that f_n converges to f_0; that is, $x \in X$ implies $f_n(x) \to f_0(x)$. The notation
$f_n \rightrightarrows f_0$ means that f_n *converges uniformly to* f_0; that is, if $\bar{\rho}\{f_n, \ f_0\} =$
$sup \ \rho\{f_n(x), \ f_0(x)\}$, $x \in X$, where ρ is the distance function in Y, then
$\bar{\rho}\{f_n, \ f_0\} \to 0$.

<u>1.2</u>   With this background, suppose $\mathfrak{X}$ is the class of Peano spaces (that
is, the totality of spaces each of which is the image, under some map-

ping, of a closed interval; see 4.1) while $\mathfrak{I}$ is the class of mappings $f: X \to Y$, where $X \in \mathfrak{X}$ and Y is metric.

The fundamental concept is a restatement of the condition that the Fréchet distance between two mappings is zero. A mapping $f_1: X \to Y$ is said to be *Fréchet equivalent* to a mapping $f_2: X_2 \to Y$ (notation: $f_1 \sim f_2$) if and only if for every $\epsilon > 0$ there is a homeomorphism $h_\epsilon: X_1 \approx X_2$ such that:

$$\overline{\rho}\{f_1, f_2 h_\epsilon\} < \epsilon.$$

In considering this concept, let it be understood that the letter h (with or without subscript) indicates a homeomorphism. Then $f_1 \sim f_2$ if and only if for every $\epsilon > 0$ the following situation exists:

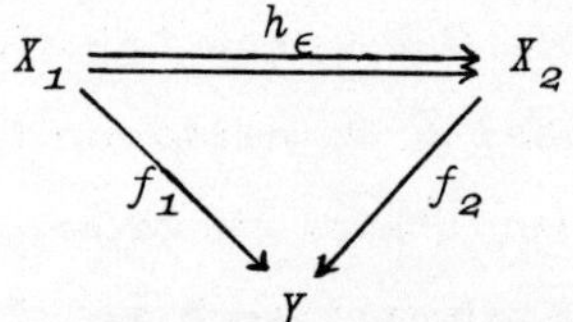

$$\overline{\rho}\{f_1, f_2 h_\epsilon\} < \epsilon.$$

It is to be noted that $f_1(X)$ and $f_2(X)$ must be in the same metric space before the mappings have any chance of being Fréchet equivalent. The diagram shows that beginning with *any* $x \in X$, one can go into Y by means of f_1, or by h_ϵ followed by f_2, and the images of x in Y are within ϵ of each other. Under these circumstances, it will be stated that there is an *approximate matching* between the mappings f_1 and f_2. In the event $\overline{\rho}\{f_1, f_2 h_\epsilon\} = 0$ for some h_ϵ, then the diagram is said to be *commutative* ($f_1 = f_2 h_\epsilon$) and the *matching* is said to be *exact*.

It is easy to see that an equivalent definition is the following: $f_1 \sim f_2$ if and only if there is a sequence, $\{h_n\}$, of homeomorphisms, $h_n: X_1 \approx X_2$, $n = 1, 2, 3, \cdots$, such that $f_2 h_n \rightrightarrows f_1$.

The relation $\sim$ is readily seen to be an equivalence relation over the class $\mathfrak{I}$ and so partitions it into mutually exclusive equivalence classes, $[f]$. Each equivalence class, $[f]$, is known as a *Fréchet variety*, **V**. Any representative of the equivalence class — that is, any mapping in $[f]$ — is said to be a *representation* of the Fréchet variety, **V**.

It is now possible to state the *representation problem for Fréchet varieties:*

Given one representation of a Fréchet variety, find all of its representations.

1.3   Of course, an obvious solution is this: If f is a representation of V, then the totality of representations of V is the totality of mappings, g, such that $f \sim g$. On the other hand, the definition of Fréchet equivalence is somewhat descriptive and often difficult to use, so that what one really desires is an equivalent definition which is more constructive in character. The principal reason for the difficulty encountered in the use of the notion of Fréchet equivalence, by the way, is that the definition is in terms of an *approximate* rather than an *exact matching* (see 1.2). On the other hand, the analytic theory of surfaces makes this definition of an equivalence practically mandatory (see Youngs [14]).

The importance of the problem is due principally to the fact that a Fréchet variety, known in terms of a particular representation may have, for analytic or other reasons, more favorable representations. Or as Radó [8, p.420] has put it, given a particular mapping, f, there may be more favorable mappings, g, in the collection of solutions of the relation $f \sim g$. The representation problem, therefore, asks for suitable criteria with which to test the validity of the statement $f \sim g$; any such criterion will be called an F—criterion, (see Radó [8, p.420]).

1.4   A first simple attack on the problem might attempt to capitalize directly on the fact that if two mappings are Fréchet equivalent, then it is possible to match them approximately, the degree of approximation being entirely within one's control. It is not unnatural to feel that, since the accuracy of the approximation can be controlled, an exact matching should be possible, thus arriving at a simple F—criterion. That is, if $f_1 \sim f_2$ it should be possible, one might feel, to find a homeomorphism $h : X_1 \approx X_2$ such that $f_1 = f_2 h$. This is not the case, as the following example shows.

Suppose that $X_1 = X_2 = X$ is the set of points $0 \leq x \leq 1$ on the real line. The mappings f_1 and f_2 will be defined so as to carry X onto itself in the following manner:

$$f_1(x) = \begin{cases} 0 & \text{if } 0 \leqq x \leqq 1/2. \\[2ex] 2x - 1 & \text{if } 1/2 \leqq x \leqq 1. \end{cases}$$

$$f_2(x) = x.$$

Now for any $\epsilon > 0$, let

$$h_\epsilon(x) = \begin{cases} 2\epsilon x & \text{if } 0 \leqq x \leqq 1/2. \\[2ex] 2(1 - \epsilon)x - (1 - 2\epsilon) & \text{if } 1/2 \leqq x \leqq 1, \end{cases}$$

Then:

$$\rho\{f_1(x),\ f_2 h_\epsilon(x)\} = \begin{cases} 2\epsilon x & \text{if } 0 \leqq x \leqq 1/2. \\[2ex] 2\epsilon(1 - x) & \text{if } 1/2 \leqq x \leqq 1. \end{cases}$$

Therefore:

$$\overline{\rho}\{f_1,\ f_2 h_\epsilon\} < \epsilon, \quad \text{and} \quad f_1 \sim f_2.$$

On the other hand, it is quite obvious that any attempt to obtain a homeomorphism $h: X_1 \approx X_2$ such that $f_1 = f_2 h$ is foredoomed to failure because f_1 maps the segment $0 \leqq x \leqq 1/2$ onto a single point, while $f_2 h$ is topological.

This simple example shows that *approximate matching is an essential feature of Fréchet equivalence,* and the obvious attack on the problem fails. This being the case, it is well to consider a few simple necessary conditions which result as a consequence of the hypothesis that $f_1 \sim f_2$ in the hope that a large body of such necessary conditions will supply a sufficient condition.

1.5   *A first necessary condition is that the range spaces X_1 and X_2 be homeomorphic.*

A second necessary condition is that $f_1(X_1) = f_2(X_2)$.

Both conditions are immediate consequences of the definition of Fréchet equivalence (see 1.2), and show, by the way, two highly important properties of Fréchet varieties. As these properties are often misunderstood, a few comments will not be out of place.

First, if V is a Fréchet variety with representations $f_1: X_1 \to Y$ and $f_2: X_2 \to Y$, then the fact that X_1 and X_2 are topologically equivalent makes it possible to associate, with V, a class, $[V]$, of Peano spaces, V,

each of which is homeomorphic to the range space of any representation of V.
Moreover, it is now possible to catalog Fréchet varieties, V, in terms of
the associated classes, $[V]$. For example, a Fréchet variety, V, is called
a Fréchet *curve* if and only if $[V]$ is the class of 1-cells (= closed arcs)
or 1-spheres (= Jordan curves). Indeed, subclassifications are possible, for
if $[V]$ is the class of 1-cells, then the Fréchet curve, V, is known as a
Fréchet curve *of the type of a* 1-cell; if $[V]$ is the class of 1-spheres,
then the Fréchet curve, V, is known as a Fréchet curve *of the type of a*
1-sphere.

A Fréchet variety, V, is called a Fréchet *surface* if and only if
$[V]$ is in the class of 2-manifolds (= compact connected 2-manifolds with or
without boundary). Consequently, in the spirit of further subclassification,
if $[V]$ is the class of 2-spheres, the Fréchet surface, V, is known as a
Fréchet surface *of the type of a* 2-sphere; if $[V]$ is the class of 2-cells,
the Fréchet surface is known as a Fréchet surface *of the type of a* 2-cell; etc.

In general, a Fréchet variety, V, is called a Fréchet *n-manifold*
if and only if some space in $[V]$ is an n-manifold. (This classification is
certainly not exhaustive, as there are Peano spaces which are not manifolds.)

1.6   So much for what can be done with the first necessary condition. A
further property results as a consequence of the second necessary condition,
namely, that $f_1(X_1) = f_2(X_2)$. This shows that the image of the range space
is independent of the representation. Hence with each Fréchet variety, V,
there is associated a particular Peano space, V^*, which is precisely $f(X)$
for any representation, $f : X \to Y$, of V. This Peano space, V^*, associated
with the Fréchet variety, V, is not to be confused with the variety itself—
V^* is a Peano space, while V is an equivalence class of mappings. In fact,
a source of some misunderstanding is the fact that if one is given two Peano
spaces, X and Y, the former being non-degenerate, then there is a Fréchet
variety, V, such that $[V]$ is the class of spaces homeomorphic to X, and
$V^* = Y$. In other words, there is a mapping $f : X \rightrightarrows Y$ (see Whyburn [10, p. 34,
Theorem 4.6]). Consequently, it is possible, to take a simple example, for V
to be a Fréchet curve and for V^* to be a 2-sphere — a situation which might
tempt one to call V a surface. This shows that knowing the space V^* is of
absolutely no help in classifying the Fréchet variety V as to type; it is

essential to know $[V]$.

<u>1.7</u>   But to return to the main body of the argument, it is not surprising that the two obvious necessary conditions of 1.5 lack sufficiency.   For example, suppose $X_1 = X_2 = X$ is the set of points $|z| \leqq 1$ on the complex z-plane. Let $f_1(z) = z$ and $f_2(z) = z^2$.   Then $f_1(X) = f_2(X)$, but the mappings are clearly not Fréchet equivalent.

These necessary conditions are, of course, on a very easy level.   A much more sophisticated necessary condition, and the first non-trivial condition to be known, was discovered by Kerékjártó [5] in 1926 for Fréchet surfaces of the type of the 2-sphere. (Kerékjártó's argument is readily generalized to cover all Fréchet varieties, see Youngs [17]; in fact, Radó [8] has recently used a new approach for this purpose which is the ultimate in simplicity, and the reader should consult his paper on this point.)

Before stating the Kerékjártó condition it is well, from the point of view of motivation, to look at the example of 1.4 showing that exact matching is usually impossible.   This example shows precisely where the difficulty lies; it is because one of the mappings takes a non-degenerate continuum into a single point.   Indeed, if $f_1 \sim f_2$ and neither f_1 nor f_2 maps any non-degenerate continuum into a single point — such mappings are called *light* (see 3.3) — then it can be shown that an exact matching is possible, as is now well known and will be shown below.   It was Kerékjártó who first observed this fortunate property for light mappings from 2-spheres. He also used the fact that any mapping, $f : X \to Y$, can be *factored* and written as the product, or composition, of two mappings, the first *monotone* (see 3.2) and the second *light*.

Specifically, if $f : X \to Y$ is a mapping, then there is a monotone mapping $m : X \rightrightarrows \mathfrak{X}$ and a light mapping $l : \mathfrak{X} \to Y$ such that $f = lm;$ that is, $f(x) = l(m(x))$ for every $x \in X$ (see the Eilenberg–Whyburn Factor Theorem, 3.5). (The letters m and μ will invariably be used for monotone mappings, l and λ for light.)   The space $\mathfrak{X}$ is called the *middle space* of the monotone light factorization, lm, of f.   The situation can be represented in a commutative diagram:

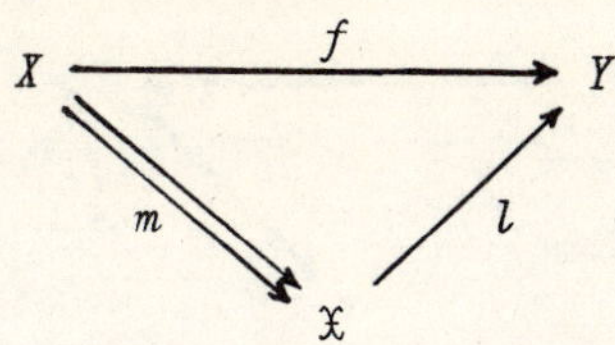

The factorization is not, strictly speaking, unique — more will be said of this later. (Double arrows will be omitted from diagrams in the future.)

Now suppose that $f_1 : X_1 \to Y$ and $f_2 : X_2 \to Y$ are Fréchet equivalent, while $l_1 m_1$ and $l_2 m_2$ are monotone-light factorizations of f_1 and f_2 with middle spaces $\mathfrak{X}_1$ and $\mathfrak{X}_2$, respectively. Having successfully completed a similar body of research in case X_1 and X_2 are 1-spheres, Kerékjártó sought to show that there was an *exact matching of the light factors* of the two mappings. In this endeavor Kerékjártó was successful; specifically, he was able to show that there is a homeomorphism $h : \mathfrak{X}_1 \approx \mathfrak{X}_2$ such that $l_1 = l_2 h$. To be quite precise, Kerékjártó showed this to be the case if X_1 and X_2 are both 1-spheres or both 2-spheres. In fact, this is true in ultimate generality within the class of Peano spaces, as was indicated at the beginning of 1.7.

One can now prove the following statement. *If $f_1 : X_1 \to Y$ and $f_2 : X_2 \to Y$ are two light mappings which are Fréchet equivalent, then there is a homeomorphism $h : X_1 \approx X_2$ such that $f_1 = f_2 h$.*

The proof uses the fact that the identity mapping $k_i : X_i \rightrightarrows X_i$ (defined by the formula $k_i(x_i) = x_i$, $x_i \in X_i$) is monotone, $i = 1, 2$. Hence $f_i k_i$ is a monotone-light factorization of f_i with middle space X_i, $i = 1, 2$. The result follows on employing the Kerékjártó necessary condition.

The Kerékjártó condition was the first non-trivial necessary condition for Fréchet equivalence and may be re-stated in a highly convenient form. If $f_1 \sim f_2$, then there are monotone-light factorizations, $f_1 = l m_1$ and $f_2 = l m_2$, with the same middle space. (The mapping l here is l_2 above, while the mapping m_1 is $h m_1$.) In other words, if for every $\epsilon > 0$ there is a homeomorphism h_ϵ such that:

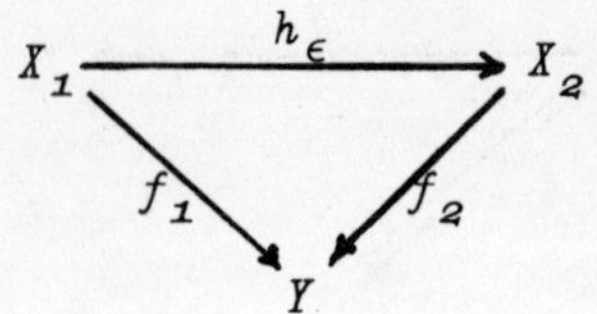

$$\overline{\rho}\{f_1,\ f_2 h_\epsilon\} < \epsilon,$$

then one has the following commutative diagram:

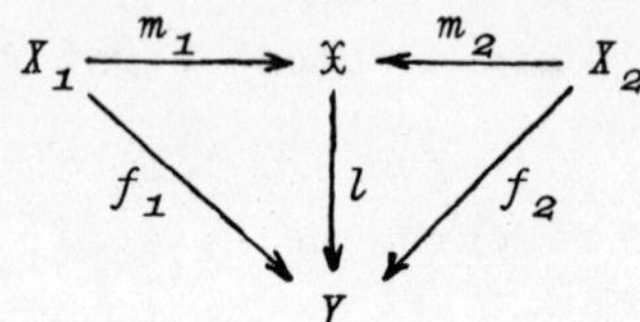

On the other hand, this is *not a sufficient condition,* as was originally supposed (for a history of this error the reader may consult Youngs [17]), but all the known sufficient conditions are arrived at by suitable refinements of this necessary condition of Kerékjártó. Indeed, this necessary condition may be said to exhibit *a standard for desirable F-criteria.* First, an *exact matching* at the level of the range spaces is usually impossible, but the necessary condition of Kerékjártó shows that *exact matching* at the level of a middle space is possible after the mappings have been suitably factored. Second, from the level of the middle spaces on it is highly important, for analytic reasons, to have the matched mappings *light.*

1.8   To return to the necessary condition of Kerékjártó, it is now easy to see how it can be modified to obtain a certain type of sufficient condition. The salient remark is this: $f_1 \sim f_2$ if there are monotone–light factorizations, $l m_1$ for f_1 and $l m_2$ for f_2 such that $m_1 \sim m_2$. (See Youngs [11].)

Of course, this sufficient condition merely pushes back the problem of approximately matching the mappings themselves to the problem of approximately matching monotone factors of the mappings, and so is subject to some, but by no means all, of the same difficulties as the original definition of Fréchet equivalence. On the other hand, it is important to remark that this sufficient condition is also necessary (see Youngs [11]), and so reduces the problem to finding F-criteria for monotone mappings rather than general mappings, a simplification which is in a very real sense the key to this paper. In view of this fact, the complete statement of this simplification is explicitly displayed below.

REDUCTION THEOREM: *If X_1 and X_2 are Peano spaces, then two mappings, $f_1:X_1 \to Y$ and $f_2:X_2 \to Y$, are Fréchet equivalent if and only if there are monotone–light factorizations, lm_1 for f_1 and lm_2 for f_2, such that m_1 and m_2 are Fréchet equivalent.*

An immediate consequence of this reduction theorem, by the way, is that, much after the character of 1.6, a class of spaces can be associated with a Fréchet variety. Specifically, if V is a Fréchet variety, f is a representation of V, and lm is a monotone–light factorization of f with middle space $\mathfrak{B}$, then let $[\mathfrak{B}]$ be the class of spaces homeomorphic to $\mathfrak{B}$. It is easy to see that, for any space in $[\mathfrak{B}]$ there is a monotone–light factorization of some representation of V which has this space as its middle space. Conversely, any monotone–light factorization of any representation of V has a middle space in $[\mathfrak{B}]$.

2. The representation problem for Fréchet surfaces

<u>2.1</u>   The discussion up to this point has been on the level of complete generality within the meaning of Fréchet varieties, and it has been possible to arrive at a reduction theorem. To progress to an F–criterion of general usefulness, one can now work on the problem of determining necessary and sufficient conditions for two *monotone* mappings from general Peano spaces to be Fréchet equivalent. A solution of this problem appears to be completely beyond the present mathematical horizon. At the moment there is every reason to consider that there is no hope of an F–criterion for a Fréchet variety V unless the spaces in $[\mathfrak{B}]$ have a relatively simple structure; in other words, unless the spaces in $[V]$ are almost trivially simple. (This opinion is based, of course, on the classical plan of attack, and it is entirely possible that a new approach may invalidate it.) On the other hand, it should be noted that the contemplated applications are of such a character that the spaces in $[V]$ are indeed simple, being no more complicated than 2–manifolds.

The simplest case occurs if $[V]$ is the class of arcs [Jordan curves], for then any space in $[\mathfrak{B}]$ is an arc or a single point [a Jordan curve or a single point]. One might expect that an F—criterion for such Fréchet varieties, that is, for Fréchet curves, is readily obtainable, and this is the case, since *the necessary condition of Kerékjártó is easily seen to be sufficient.*

If $[V]$ is the class of 2—spheres, then any space in $[\mathfrak{B}]$ is a cactoid (see 6.1), and the situation is at once immeasurably complicated. An F—criterion, however, has been given for this case by Youngs [17].

If $[V]$ is the class of spaces homeomorphic to a given 2—manifold V, then the possible monotone images of V are known, thanks to a paper by Roberts and Steenrod [9], but are far beyond cactoids in complexity. On the other hand, if $[V]$ is the class of spaces homeomorphic to a given 3—cell V, then the possible monotone images of V have not been determined but certainly cover an immense class of Peano spaces; in fact, it rather appears that before one can do anything on this case (along the lines of this paper) a characterization of the 3—cell is required. This perhaps serves to explain the rather pessimistic comments above.

It appears fairly obvious, therefore, that if the complexity of $[\mathfrak{B}]$ is, as every indication suggests, the determining factor in the difficulty of finding an F—criterion, there is very little hope of solving the representation problem for general Fréchet varieties.

2.2   The successful attack on the case in which $[V]$ is the class of 2—spheres suggests the possibility of obtaining a solution for any class, $[V]$, consisting of 2—manifolds. In other words, perhaps the representation problem can be solved for *all Fréchet surfaces* (see 1.2). In view of the reduction theorem, one can consider two monotone mappings,

$$m_1 \colon X_1 \rightrightarrows \mathfrak{X} \quad \text{and} \quad m_2 \colon X_2 \rightrightarrows \mathfrak{X},$$

where X_1 and X_2 are homeomorphic 2—manifolds, and ask for conditions which are both necessary and sufficient for

$$m_1 \sim m_2.$$

2.3   In the event the 2—manifolds have boundaries this introduces a certain lack of homogeneity into the situation and one can expect complications which will not arise if the 2—manifolds are closed. A digest of the manner in which the boundaries are handled will follow before considering the smooth case for closed 2-manifolds.

2.4   With the conventions of 2.2, suppose X_1 and X_2 are 2-cells while $\mathfrak{X}$ is the simplest possible non-degenerate Peano space, an arc. An example of Radó [8, p. 438] shows one cannot assert that $m_1 \sim m_2$.

The key to the counter-example is the fact that $\dot{X}_1$, the bounding curve of X_1, may be mapped by m_1 onto one end point of the arc $\mathfrak{X}$, while $\dot{X}_2$ may be mapped by m_2 onto the other end point. Since the homeomorphisms demanded for Fréchet equivalence must take $\dot{X}_1$ onto $\dot{X}_2$, it is clear that $m_1 \sim m_2$ is false.

2.5   With the conventions of 2.4 suppose $\mathfrak{X}$ is a 2-sphere. Here too it is impossible to assert that $m_1 \sim m_2$ and again the counter-example is due to Radó [8, p. 439].

The Jordan curve $\dot{X}_1$ is mapped by m_1 onto a single point of $\mathfrak{X}$, and $\dot{X}_2$ is mapped by m_2 onto any other point of $\mathfrak{X}$, the mappings being homeomorphisms elsewhere. It is clear, for the same reason as indicated in the example of 2.4, that $m_1 \sim m_2$ is false.

2.6   Further examples of this type will occur to the reader but these should suffice for the motivation of the discussion to follow.

If H is a 2-manifold with boundary it is possible to cap the boundary curves with 2-cells so as to obtain a closed 2-manifold X. Suppose that the capping 2-cells are $C_1, \cdots, C_n$; that is, the boundary curves of H are $\dot{C}_1, \cdots, \dot{C}_n$, where $\dot{C}_i$ is the bounding 1-sphere of C_i, $i = 1, \cdots, n$.

If $m : H \rightrightarrows \mathfrak{H}$ is a monotone mapping then it is possible, roughly speaking, to adjoin an open 2-cell $\mathfrak{R}_i$ to $m(\dot{C}_i)$, $i = 1, \cdots, n$ in such a fashion that:

$$\mathfrak{H} \cup \bigcup \mathfrak{R}_i \text{ is a Peano space } \mathfrak{X},$$
$$\text{The frontier of } \mathfrak{R}_i \text{ is } m(\dot{C}_i), \quad i = 1, \cdots, n,$$
$$R_i \cap R_j = 0; \quad i \neq j; \quad i, j = 1, \cdots, n.$$

This is easily seen to be the case in the examples of 2.4 and 2.5; it is proved in general, using the theory of upper semi-continuous collections, in 3.5. Moreover, it can be shown that there is a mapping

$$\mu : X \rightrightarrows \mathfrak{X}$$

such that if $(\mu|H)(x) = \mu(x)$, $x \in H$, then

$$\mu\,|\,H \,=\, m,$$
$$x \,=\, \mu^{-1}\mu(x), \quad x \in X - H.$$

2.7 Now consider two monotone mappings

$$m_i:H_i \rightrightarrows \mathfrak{H}, \quad i = 1,\ 2$$

with the understanding that H_1 and H_2 are homeomorphic 2–manifolds with boundary.

Suppose X_i is a closed 2–manifold obtained by capping the boundary curves of H_i with 2–cells $C_{i\,1}, \cdots, C_{in}$, while $\mathfrak{X}_i$ is the Peano space obtained as in 2.6 by adjoining open 2–cells to $\mathfrak{H}$, and μ_i is the mapping of 2.6, $i = 1,\ 2$.

In neither of the examples considered in 2.4 and 2.5 is it true that $\mathfrak{X}_1 = \mathfrak{X}_2$.

On the other hand, it can be shown that *if* $m_1 \sim m_2$ *then, roughly speaking,* $\mathfrak{H}$ *can be enlarged to* $\mathfrak{X}_1$ *and* $\mathfrak{X}_2$ *in such a fashion that* $\mathfrak{X}_1 = \mathfrak{X}_2$ *and, in addition,* $\mu_1 \sim \mu_2$. (See 15.6.)

The assertion that the above condition is *sufficient* to guarantee that $m_1 \sim m_2$ may occasion no astonishment, yet no simple direct proof of this fact is available, and it is one of the concluding results of this paper (see 23.5).

For the purposes of the present discussion, however, these comments should suffice to show that *the consideration of F–criteria for monotone mappings from 2–manifolds may be restricted to the case in which the mappings are from closed 2–manifolds.*

2.8 In consequence of the remarks of 2.7 from this point on it will be assumed that the monotone mappings

$$m_1:X_1 \rightrightarrows \mathfrak{X} \quad \text{and} \quad m_2:X_2 \rightrightarrows \mathfrak{X}$$

of 2.2 are from homeomorphic closed 2–manifolds X_1 and X_2.

It has been observed that even in the event X_1 and X_2 are 2–spheres, there is no guarantee that $m_1 \sim m_2$. (See Youngs [17].) In the search for a suitable additional condition one now turns to algebraic topology.

2.9 The mapping $m_i:X_i \rightrightarrows \mathfrak{X}$ induces a homomorphism $m_i^*:H^2(\mathfrak{X}) \rightarrow H^2(X_i)$, $i = 1,\ 2$, where the groups involved are the 2–dimensional Čech cohomology groups with the additive group of integers taken as the coefficient group. (See 7.1 – 7.9.) If $m_1 \sim m_2$ then there is a sequence $\{h_n\}$ of homeomorphisms $h_n : X_1 \approx X_2$ such that $m_2 h_n \rightrightarrows m_1$. Since the group $H^2(X_i)$ is cyclic infinite if X_i is orientable, and of order 2 otherwise, $i = 1, 2,$ while the induced homomorphisms h_n^* are isomorphisms from $H^2(X_2)$ onto $H^2(X_1)$, $n = 1, 2, 3, \cdots$, it may be assumed that h_n^* is independent of n. Let $\eta = h_n^*$. It is well known that $(m_2 h_n)^*$ converges to m_1^* and since $(m_2 h_n)^* = h_n^* m_2^* = \eta m_2^*$, this means that $\eta m_2^* = m_1^*$.

Hence $m_1 \sim m_2$ implies that there is an isomorphism $\eta : H^2(X_2) \approx H^2(X_1)$ such that commutativity holds in the following diagram:

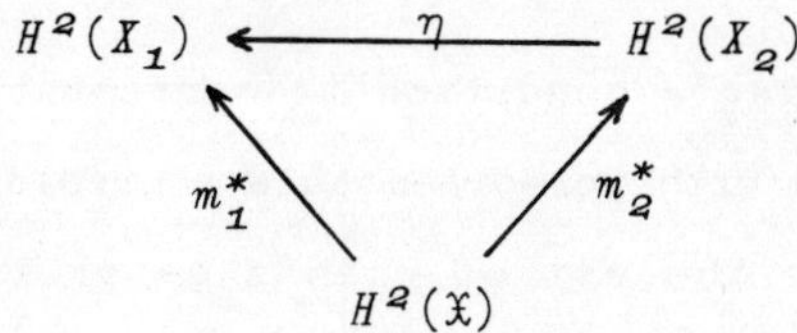

One way of interpreting this result along the lines of 1.2 is to say that arbitrarily close approximate matching on the part of m_1 and m_2 implies exact matching of m_1^* and m_2^* — arbitrarily close approximate matching at the set theoretic level implies exact matching at the algebraic level. On comparing this situation with the concluding remarks of 1.7, it will be observed that this fits well with the aims there enunciated.

In the event X_1 and X_2 are 2–spheres, it has been shown that exact matching of m_1^* and m_2^* is sufficient to guarantee that $m_1 \sim m_2$. (See Youngs [17] together with Borsuk [1] and Youngs [15].) Thus in the 2–sphere case $m_1 \sim m_2$ *if and only if exact matching of* m_1^* *and* m_2^* *is possible.*

From the aesthetic point of view this is a rather satisfactory F-criterion. Unfortunately, it breaks down if X_1 and X_2 are not 2–spheres.

2.10   With the conventions of 2.8, suppose $X_1 = X_2 = X$, a torus, $\mathfrak{X}$ is a 2–sphere, and there is an isomorphism $\eta : H^2(X) \approx H^2(X)$ such that $m_1^* = \eta m_2^*$. It is not possible to assert that $m_1 \sim m_2$.

Let G be the interior of a 2–cell on X while $\mathfrak{x}_1$ and $\mathfrak{x}_2$ are distinct points of $\mathfrak{X}$. There is certainly a homeomorphism $\mathfrak{h} : \mathfrak{X} \approx \mathfrak{X}$ such that $\mathfrak{h}(\mathfrak{x}_1) = \mathfrak{x}_2$ and $\mathfrak{h}^* : H^2(\mathfrak{X}) \approx H^2(\mathfrak{X})$ is the identity; that is, $\mathfrak{h}^*$ maps each element of $H^2(\mathfrak{X})$ into itself.

There is a mapping $m_1 : X \rightrightarrows \mathfrak{X}$ which is a homeomorphism on G and such that $m_1(X - G) = \mathfrak{x}_1$. Define $m_2 = \mathfrak{h}m_1$. Now $m_2^* = (\mathfrak{h}m_1)^* = m_1^*\mathfrak{h}^* = m_1^*$, and the identity isomorphism $\eta : H^2(X) \approx H^2(X)$ provides an exact matching of m_1^* and m_2^*.

Nevertheless, m_1 and m_2 are not Fréchet equivalent, a statement which follows readily from the fact that $\mathfrak{x}_1 \neq \mathfrak{x}_2$.

The root of the trouble is a point–set theoretic difficulty. If $\mathfrak{C}$ is a 2–cell on $\mathfrak{X}$ and $\mathfrak{C}^o$ its interior, then it is not true that $m_1^{-1}(\mathfrak{C}^o) \approx m_2^{-1}(\mathfrak{C}^o)$ as would have been the case had X_1 and X_2 been 2–spheres.

Before suggesting a condition to correct this matter, first consider an example concerned with non–orientable manifolds.

2.11   Suppose X is the set of points determined by the inequality $|z| \leqq 1$ on a complex z–plane with the understanding that diametrically opposite points on $\dot{X}$, the set of points determined by $|z| = 1$, are identified. In short, X is a projective plane. It will be understood that $X_1 = X = X_2$.

The space $\mathfrak{X}$ is a pair of tangent geometric 2–spheres $\mathfrak{X}_1$ and $\mathfrak{X}_2$ with $\mathfrak{X}_1 \cap \mathfrak{X}_2 = \mathfrak{x}$. Let $\overline{\mathfrak{x}}$ be the point diametrically opposite to $\mathfrak{x}$ on $\mathfrak{X}_2$. There is a homeomorphism $\mathfrak{h} : \mathfrak{X} \approx \mathfrak{X}$ such that $\mathfrak{h}$ is the identity on $\mathfrak{X}_1$, and on $\mathfrak{X}_2$ it is a reflection about a great circle of $\mathfrak{X}_2$ passing through $\mathfrak{x}$.

The mapping m_1 is defined to take X homeomorphically onto $\mathfrak{X}$ except that the 1–spheres determined by $|z| = 1/2$ and $|z| = 1$ are mapped onto $\mathfrak{x}$ and $\overline{\mathfrak{x}}$ respectively. Define $m_2 = \mathfrak{h}m_1$.

The group $H^2(\mathfrak{X})$ is the direct sum of groups W_1 and W_2 isomorphic to $H^2(\mathfrak{X}_1)$ and $H^2(\mathfrak{X}_2)$ respectively; that is

$$H^2(X) = W_1 \oplus W_2.$$

If w_i is a generator for the cyclic infinite group W_i, $i = 1, 2$, then it follows from the definition of $\mathfrak{h}$ that for any integers a and b

$$\mathfrak{h}^*(aw_1 + bw_2) = aw_1 - bw_2.$$

Consequently:

$$m_2^*(aw_1 + bw_2) = (\mathfrak{h}m_1)^*(aw_1 + bw_2)$$
$$= m_1^*\mathfrak{h}^*(aw_1 + bw_2)$$

$$= m_1^*(aw_1 - bw_2)$$

$$= m_1^*(aw_1) - m_1^*(bw_2)$$

$$= m_1^*(aw_1) + m_1^*(bw_2), \text{ since } H^2(X) \text{ is of order 2.}$$

$$= m_1^*(aw_1 + bw_2)$$

Therefore the identity isomorphism $\eta: H^2(X) \approx H^2(X)$ provides an exact matching of m_1^* and m_2^*. Moreover this example does not have the point-set theoretic difficulties found in 2.10.

Nevertheless, it is not true that $m_1 \sim m_2$. (Compare this example with the counter-example of Youngs [17] in the 2-sphere case.)

To achieve a rough understanding of the situation, suppose $\mathfrak{U}$ is a region (= connected open set) consisting of the union of the interiors, $\mathfrak{U}_1$ and $\mathfrak{U}_2$, of two 2-cells in X having the property that $\mathfrak{U}_i \subset X_i$ and $\mathfrak{h}(\mathfrak{U}_i) = \mathfrak{U}_i$, $i = 1, 2$. In view of these facts, $x \in \mathfrak{U}$, and $U_1 \equiv m_1^{-1}(\mathfrak{U}) = m_2^{-1}(\mathfrak{U}) \equiv U_2$. Let $U = U_1 = U_2$. If arbitrarily close matching of m_1 and m_2 is possible, then arbitrarily close matching of $\underline{m}_1$ and $\underline{m}_2$ can be achieved, where $\underline{m}_i = m_i | U$, $i = 1, 2$. Using the locally compact cohomology theory (See 8.1 to 8.7) this implies, as in 2.9, that there is an isomorphism $\eta: H^2(U) \approx H^2(U)$ such that $\underline{m}_1^* = \eta \underline{m}_2^*$ where $\underline{m}_i^*: H^2(\mathfrak{U}) \to H^2(U)$ is the homomorphism induced by $\underline{m}_i: U \to \mathfrak{U}$, $i = 1, 2$. (Notice that $H^2(U)$ is cyclic infinite.)

But $H^2(\mathfrak{U}) = W_1 \oplus W_2$, where $W_i \approx H^2(\mathfrak{U}_i)$, $i = 1, 2$. Let w_i be a generator of the cyclic infinite group W_i, $i = 1, 2$. If $\underline{\mathfrak{h}} = \mathfrak{h} | \mathfrak{U}$, then

$$\underline{m}_2^*(w_1) = \underline{m}_1^* \underline{\mathfrak{h}}^*(w_1) = \underline{m}_1^*(w_1),$$

while

$$\underline{m}_2^*(w_2) = \underline{m}_1^* \underline{\mathfrak{h}}^*(w_2) = \underline{m}_1^*(-w_2) = -\underline{m}_1^*(w_2).$$

Hence a matching isomorphism $\eta: H^2(U) \approx H^2(U)$ exists if and only if $\underline{m}_1^*$ and $\underline{m}_2^*$ are trivial; that is $\underline{m}_i^*(H^2(\mathfrak{U})) = 0$. But it can be shown that neither of these homomorphisms is trivial.

<u>2.12</u>   A condition suggested by the discussion of these last two examples is now to be developed using cyclic element theory (see Whyburn [10]).

If X is the image, under a monotone mapping, of a closed 2-manifold I, then X is to be called a *mantoid*. (If I is a 2-sphere, then X is a *cactoid*.)

A set $\mathfrak{U}$ in a mantoid $\mathfrak{X}$ is said to be a *normal region* if and only if:

1° $\mathfrak{U}$ is a non-vacuous, connected, open set.

2° $\dot{\mathfrak{U}}$, the frontier of $\mathfrak{U}$ in $\mathfrak{X}$, has a finite number of components, each a point or a Jordan curve. (In consequence, it is easy to show that $\bar{\mathfrak{U}}$ is a Peano space.)

3° If x is a point component of $\dot{\mathfrak{U}}$, then it is an end point of $\bar{\mathfrak{U}}$.

4° If $\mathfrak{J}$ is a Jordan curve component of $\dot{\mathfrak{U}}$, then the true cyclic element $\mathfrak{C}$ of $\bar{\mathfrak{U}}$ which contains $\mathfrak{J}$ is a 2-cell. (It follows that $\dot{\mathfrak{C}} = \mathfrak{J}$, where $\dot{\mathfrak{C}}$ is the Jordan curve bounding $\mathfrak{C}$.)

With these preparatory remarks, it is now possible to define a condition to be denoted by C.

Definition: Two monotone mappings, $m_1 : X_1 \rightrightarrows \mathfrak{X}$ and $m_2 : X_2 \rightrightarrows \mathfrak{X}$, where X_1 and X_2 are closed 2-manifolds, are said to satisfy condition C if and only if, for each normal region $\mathfrak{U}$ in $\mathfrak{X}$, setting $U_i = m_i^{-1}(\mathfrak{U})$ and $\underline{m}_i = m_i | U_i$, $i = 1, 2$, there is a homeomorphism $h : U_1 \approx U_2$ such that commutativity holds in the following diagram:

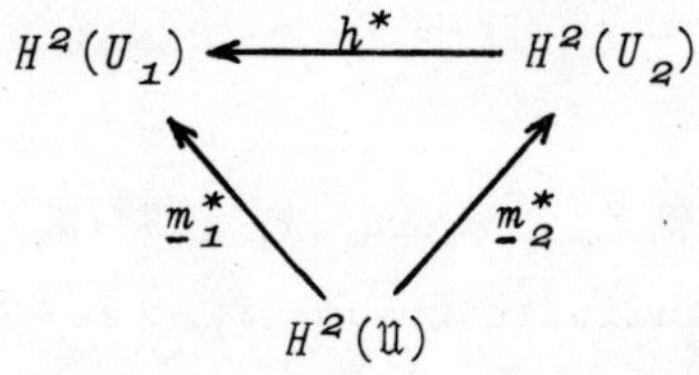

$$H^2(U_1) \xleftarrow{\quad h^* \quad} H^2(U_2)$$
$$\underline{m}_1^* \nwarrow \qquad \nearrow \underline{m}_2^*$$
$$H^2(\mathfrak{U})$$

Notice that both of the troublesome features noticed in the preceding examples are taken care of by this condition: the difficulties of the first example are eliminated by requiring the existence of the homeomorphism h, the difficulties of the second by requiring that h^* be a matching isomorphism.

One of the basic results in this paper follows.

THEOREM: *If $m_1 : X_1 \rightarrow \mathfrak{X}$ and $m_2 : X_2 \rightarrow \mathfrak{X}$ are monotone mappings, where X_1 and X_2 are closed 2-manifolds, then $m_1 \sim m_2$ if and only if the mappings satisfy condition C.*

2.13   It should be mentioned that the principal portion of this paper is planned to provide a simultaneous F-criterion for four general types   of

Fréchet surfaces, namely Fréchet surfaces of the type of

        (i)   a closed orientable 2—manifold,

        (ii)   a closed non—orientable 2—manifold,

        (iii)   an orientable 2—manifold with boundary,

        (iv)   a non—orientable 2—manifold with boundary.

The basic condition C is obviously very strong. In the body of the paper it is shown that Fréchet equivalence implies C; it is then proved that what is called *compatibility* (a condition which, at first sight, is much weaker than C) implies Fréchet equivalence. In the event the 2—manifolds X_1 and X_2 are orientable, a still weaker condition called *o—compatibility* suffices to imply Fréchet equivalence.

Since o—compatibility is weaker than compatibility, there is perhaps room for a non—intrinsic F—criterion in the non—orientable case concerned with covering orientable 2—manifolds. While this is so, it appears at the moment that the selected procedure has advantages for the contemplated applications.

CHAPTER II

TOPOLOGICAL PRELIMINARIES

The first chapter has already indicated that the machinery to be employed is drawn from the fields of analytic (Whyburn [10]) and algebraic (Lefschetz [6]) topology. Relative to the latter, both the Čech and the locally compact cohomology theory (Lefschetz groups) are employed. Additional references here are Hurewicz–Wallman [4] and Cartan [2].

This chapter consists of a brief survey of some, but by no means all, of the topological tools to be employed.

3. Monotoneity and lightness

3.1   The basic underlying concept is that of an *upper semi-continuous decomposition* (Whyburn [10, p. 123]); in this connection, the notation and theorems to be employed are summarized in Youngs [12], and for reasons of economy will not be repeated here.

3.2   *Definition:* A mapping $m: X \rightrightarrows Y$ is said to be *monotone* if and only if $y \in Y$ implies that $m^{-1}(y)$ is a *continuum;* that is, a compact and connected set.

3.3   *Definition:* A mapping $l: X \to Y$ is said to be *light* if and only if $y \in Y$ implies that $l^{-1}(y)$ is *totally disconnected;* that is, every component of $l^{-1}(y)$ is a single point if $l^{-1}(y)$ is not empty.

3.4   *Definition:* If $f: X \to Y$ is a mapping, and $f_1: X \rightrightarrows \mathfrak{X}_1$, $f_2: \mathfrak{X}_1 \rightrightarrows \mathfrak{X}_2$, $\cdots$, $f_{n-1}: \mathfrak{X}_{n-2} \rightrightarrows \mathfrak{X}_{n-1}$, $f_n: \mathfrak{X}_{n-1} \to Y$ is a collection of mappings such that $f(x) = f_n f_{n-1} \cdots f_2 f_1(x)$, $x \in X$, then the composition $f_n \cdots f_1$ is said to be a *factorization* of f with *middle spaces* $\mathfrak{X}_1, \cdots, \mathfrak{X}_{n-1}$.

3.5   One now has the following important result.

FACTOR THEOREM: (Eilenberg–Whyburn): *If H is a closed subset of a compactum X, and $f: H \to Y$ is a mapping, then:*

    1°. *There is a monotone mapping $m: X \rightrightarrows \mathfrak{X}$ such that:*

        a) *If $m(H) = \mathfrak{H}$, then $m^{-1}(\mathfrak{H}) = H$.*

        b) *If $x \in X - H$, then $x = m^{-1}m(x)$.*

2°. *There is a light mapping* $l:\mathfrak{H} \to Y$ *such that, if* $\underline{m} = m|H$,
then $f = l\underline{m}$.

This theorem is slightly more general than the classical result of
Eilenberg and Whyburn, but its proof presents no new difficulties. The ordinary form of the theorem results if $H = X$. *Under these circumstances,* lm
is called a monotone–light factorization of f. The space $\mathfrak{X}$ is the middle
space of the factorization.

Notice that if $l_1 m_1$ is a monotone–light factorization of $f:X \to Y$
with middle space $\mathfrak{X}_1$, while $h:\mathfrak{X}_1 \underset{\sim}{} \mathfrak{X}_2$ is a homeomorphism from $\mathfrak{X}_1$ onto a
space $\mathfrak{X}_2$, then $f = (l_1 h^{-1})(hm_1)$. Moreover, $(hm_1):X \rightrightarrows \mathfrak{X}_2$ is monotone,
and $(l_1 h^{-1}):\mathfrak{X}_2 \to Y$ is light. Consequently, if $m_2 = hm_1$ and $l_2 = l_1 h^{-1}$,
then $l_2 m_2$ is a monotone–light factorization of f with middle space $\mathfrak{X}_2$.
Hence the factorization is not unique in the strict sense of the word.

It is easy to show, however (see Whyburn [10, p. 142]), that if
$l_1 m_1$ and $l_2 m_2$ are monotono–light factorizations of f with middle spaces
$\mathfrak{X}_1$ and $\mathfrak{X}_2$, respectively, then there is a *unique* homeomorphism

$$h:\mathfrak{X}_1 \underset{\sim}{} \mathfrak{X}_2$$

such that

$$m_2 = hm_1$$

and

$$l_2 = l_1 h^{-1}.$$

In other words, one has commutativity in the following diagram:

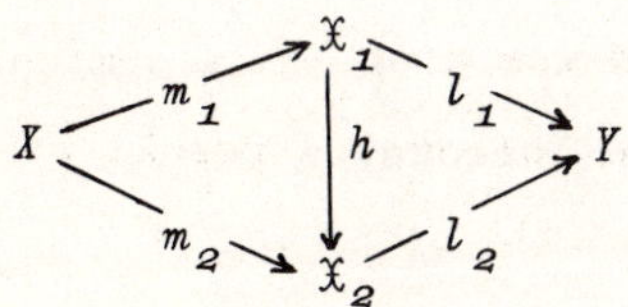

4. <u>Peano spaces</u>

From this point on it will be assumed that, unless otherwise
specified, the spaces dealt with are Peano spaces. The reader is referred
specifically to Whyburn [10, ch IV] for the general theory of Peano spaces and
the definitions of the terms involved.

<u>4.1</u>   It will be recalled that a *Peano space* is the image of a 1–cell
under a mapping into a metric space, and can be characterized as a *locally
connected continuum*. The salient point in the theory is the relatively
simple structure of a Peano space in terms of its cyclic elements — a

classical discovery of Whyburn. Certain theorems which are often employed in the sequel are stated here for convenience, it being understood that one is dealing with a Peano space, $\mathfrak{X}$. (German letters are used since in the applications Peano spaces occur as the middle spaces of monotone–light factorizations of mappings from 2–manifolds.)

Before turning to the cyclic element theory it is important to recall that:

If $\mathfrak{X}$ is a Peano space and $\mathfrak{G}$ is any open subset of $\mathfrak{X}$, then each component of $\mathfrak{G}$ is open.

If $\mathfrak{G}$ is a region (= connected open subset) of a Peano space $\mathfrak{X}$, then $\mathfrak{G}$ is arc–wise connected (= each pair of points in $\mathfrak{G}$ can be connected by an arc in $\mathfrak{G}$).

As a matter of notation, if a is an arc in $\mathfrak{X}$ with end points $\mathfrak{x}$ and $\mathfrak{y}$, then a° is defined to be the open arc $a - (\mathfrak{x} \cup \mathfrak{y})$, and $\dot{a} = \mathfrak{x} \cup \mathfrak{y}$.

__4.2__   If $\mathfrak{A}$ is an A-set of a Peano space $\mathfrak{X}$, it is important to recall that there is a *unique monotone retraction* $r_{\mathfrak{A}} \colon \mathfrak{X} \rightrightarrows \mathfrak{A}$.

If $\mathfrak{x} \in \mathfrak{X} - \mathfrak{A}$, then there is a unique component $\mathfrak{G}$ of $\mathfrak{X} - \mathfrak{A}$ containing $\mathfrak{x}$. The point $r_{\mathfrak{A}}(\mathfrak{x})$ is simply $\dot{\mathfrak{G}}$, the frontier $\mathfrak{G}$ in $\mathfrak{X}$; that is, $\overline{\mathfrak{G}} \cap \overline{(\mathfrak{X} - \mathfrak{G})}$.

Notice that
$$\rho\{r_{\mathfrak{A}}(\mathfrak{x}), \; \mathfrak{x}\} \leq max \; d(\mathfrak{G}), \; \mathfrak{G} \text{ a component of } \mathfrak{X} - \mathfrak{A}.$$

__4.3__   The theorem which follows, a result of Whyburn, is of basic importance in this paper.

CYCLIC CHAIN APPROXIMATION THEOREM: *If $\mathfrak{X}$ is a Peano space, then there is a (possibly finite) sequence, $\{\mathfrak{C}(\mathfrak{p}_k, \; q_k)\}$, of cyclic chains of $\mathfrak{X}$ such that if $\mathfrak{C}_k = \mathfrak{C}(\mathfrak{p}_k, \; q_k)$, $k = 1, 2, 3, \cdots$, then:*

$1^{\circ}.$ $\bigcup_1^n \mathfrak{C}_k \equiv \mathfrak{A}_n$ *is an A-set of $\mathfrak{X}$, $n = 1, 2, 3, \cdots$.*

$2^{\circ}.$ $\mathfrak{A}_n \cap \mathfrak{C}_{n+1} = q_{n+1}$, $n = 1, 2, 3, \cdots$.

$3^{\circ}.$ *If δ_n is the maximum of the diameters of the components of $\mathfrak{X} - \mathfrak{A}_n$, then $\delta_n \to 0$. (In the event the sequence is finite, $\delta_n = 0$ for the last value of n.)*

The collection $\{\mathfrak{C}_k\}$ is said to be a *cyclic chain approximation* to $\mathfrak{X}$. There are, in general, many different cyclic chain approximations to the same Peano space $\mathfrak{X}$.

If Ω is any *finite* set of points in $\mathfrak{X}$, then it is possible to select a cyclic chain approximation, $\{\mathfrak{C}_k\}$, of $\mathfrak{X}$ such that there is an *integer* n_o for which $n > n_o$ implies $\mathfrak{A}_n \supset \Omega$.

5. 2-manifolds

<u>5.1</u>   In this paper the term 2-cell is understood to be the homeomorphic image of the closed unit disc, $|z| \le 1$, in the complex plane. An open 2-cell is the homeomorphic image of the open disc, $|z| < 1$. Cylinder and open cylinder are similarly defined. If C is a 2-cell, then the bounding 1-sphere is designated by $\dot{C}$, and $C^o \equiv C - \dot{C}$.

<u>5.2</u>   A space X is called a closed 2-*manifold* if and only if X is a connected compactum and $x \in X$ implies that there is a neighborhood, N, of x which is an open 2-cell. A space X is a 2-*manifold with boundary* if and only if there is a closed 2-manifold Y and a finite number of 2-cells, C_1, $\cdots$, C_n such that $C_j \cap C_k = 0$, $j \ne k$; j, $k = 1$, $\cdots$, n, and $X = Y - \bigcup C_k^o$. The 1-spheres $\dot{C}_1$, $\cdots$, $\dot{C}_n$ are called the *boundary curves* of X, and by definition $\dot{X} = \bigcup \dot{C}_k$ while $X^o = X - \dot{X}$.

It should be noticed that the notation $\dot{A}$ is employed in two ways. If A is a set in a space X, then $\dot{A} = \bar{A} \cap \overline{(X - A)}$, the *frontier* of A (see 4.2); if A is a 2-manifold with boundary, then $\dot{A}$ is the union of the bounding curves. The notation A^o is sometimes employed for the interior of A in addition to being used as in 5.1 and 5.2. The context will always make the meaning clear.

<u>5.3</u>   Certain open sets on a 2-manifold are found to play a prominent role in the solution of the representation problem. These sets form a sub-class of the class of open manifolds but are designated by a name which is selected so as to avoid confusion with 2-manifolds as they are to be considered in this paper.

Definition: If X is a closed 2-manifold, U is a connected open subset of X and K is a component of $\dot{U}$ $(= \bar{U} - U)$, then U is said to *have a single cylinder of approach to* K if and only if there is an open cylinder, H, in U such that $\dot{H} = K \cup \Gamma$ (where Γ is a 1-sphere in U) and $\overline{(U - H)} \cap K = 0$. (See Roberts-Steenrod [9].)

Definition: If X is a closed 2-manifold, then U is said to be a *canonical region* of X if and only if:

      1°. U is a connected open subset of X; that is, U is a region.

      2°. $\dot{U}$ has a finite number of components, $K_1, \cdots, K_r$.

      3°. U has a single cylinder of approach to K_k, , $= 1, \cdots, r$.

An open 2-cell in a 2-manifold is a simple example of a canonical region.

Relative to these concepts, one has the following result:

THEOREM:

Given:

      1°. *A canonical region,* U, *of a closed 2-manifold* X.

      2°. *The components of* $\dot{U}$ *are* $K_1, \cdots, K_r$.

Conclusion: *There is a 2-manifold,* M, *such that:*

      3°. $U \supset M$.

      4°. $\dot{M}$ *has* r *components,* $\Gamma_1, \cdots, \Gamma_r$.

      5°. $U - M$ *has* r *components,* $H_1, \cdots, H_r$, *each an open cylinder.*

      6°. $\bar{H}_j \cap \bar{H}_k = 0, \ j \neq k; \ j, \ k = 1, \cdots, r$.

      7°. $\dot{H}_k$ *has two components, namely* K_k *and* Γ_k, $k = 1, \cdots, r$.

Proof: Let $|M|$ be the upper semi-continuous decomposition of $\bar{U}$ whose sets are the components of $\dot{U}$ and single points of U. Suppose $\phi: \bar{U} \rightrightarrows \mathfrak{M}$ is the associated mapping. Since U has a single cylinder of approach to K_k, and as ϕ is a homeomorphism on U, the point $\mathfrak{p}_k = \phi(K_k)$ has a neighborhood which is an open 2-cell, $k = 1, \cdots, r$. Consequently, $\mathfrak{M}$ is a closed 2-manifold. Select 2-cells $\mathfrak{C}_1, \cdots, \mathfrak{C}_k$ such that:

$$\mathfrak{p}_k \in \mathfrak{C}_k^{\circ}, \ k = 1, \cdots, r.$$

$$\mathfrak{C}_j \cap \mathfrak{C}_k = 0, \ j \neq k; \ j, \ k = 1, \cdots, r.$$

If $\mathfrak{N} = \mathfrak{M} - \bigcup \mathfrak{C}_k^{\circ}$, then $\mathfrak{N}$ is a 2-manifold with boundary curves $\dot{\mathfrak{C}}_1, \cdots, \dot{\mathfrak{C}}_r$. Since $x \in U$ implies $x = \phi^{-1}\phi(x)$, the set $M = \phi^{-1}(\mathfrak{N})$ is a 2-manifold with boundary $\phi^{-1}(\dot{\mathfrak{C}}_1), \cdots, \phi^{-1}(\dot{\mathfrak{C}}_r)$. The theorem follows on defining $\Gamma_k = \phi^{-1}(\dot{\mathfrak{C}}_k)$ and $H_k = \phi^{-1}(\mathfrak{C}_k^{\circ} - \mathfrak{p}_k)$, $k = 1, \cdots, r$.

Definition: If U is a canonical region of a 2-manifold, then any 2-manifold M with properties $3° - 7°$ of the theorem is said to be a 2-manifold *associated* with U.

Remark: It is easy to see that if M is a 2–manifold associated with U, then $M^° \approx U$.

6. Mantoids

Since the principal considerations in this paper will deal with mappings $m: X \rightrightarrows \mathfrak{X}$, where X is a closed 2–manifold, it is only natural that a detailed knowledge of the space $\mathfrak{X}$ will be required.

6.1   *Definition:* A space $\mathfrak{X}$ is said to be a *mantoid* if and only if it is the monotone image of a closed 2–manifold; that is to say, if and only if there is a closed 2–manifold X and a monotone mapping $m: X \rightrightarrows \mathfrak{X}$.

In particular, if X is a 2–sphere [2–cell], then $\mathfrak{X}$ is called a *cactoid* [*hemi-cactoid*]. Since a 2–manifold is clearly a locally connected continuum, it is a Peano space, hence the continuous image of an interval. Consequently, a mantoid is also the continuous image of an interval, that is, a Peano space.

It is well known that a Peano space is a cactoid if and only if each of its true cyclic elements is a 2–sphere. A similar characterization of mantoids is due to Roberts and Steenrod [9], and the next few paragraphs will be devoted to some of their results.

Definition: A Peano space is called a *generalized cactoid* if and only if each of its true cyclic elements is a closed 2–manifold and all but a finite number of the true cyclic elements are 2–spheres.

Definition: A mapping $\theta: \mathfrak{Y} \rightrightarrows \mathfrak{X}$ is said to be a *2–point identification* if and only if $\theta^{-1}(\mathfrak{x})$ is non–degenerate for exactly one point, $\mathfrak{x}_0$, in $\mathfrak{X}$, and $\theta^{-1}(\mathfrak{x}_0)$ consists of precisely two points.

Definition: A triple $(\theta, \mathfrak{Y}, \mathfrak{X})$ consisting respectively of a mapping and two Peano spaces is said to be a *suitable system* if and only if:

1° $\theta: \mathfrak{Y} \rightrightarrows \mathfrak{X}$.

2° θ is an identity mapping, or there is a factorization $\theta_k, \cdots, \theta_1$ of θ (see 3.4) such that θ_j is a 2–point identification, $j = 1, \cdots, k$.

3° $\mathfrak{Y}$ is a generalized cactoid.

If $(\theta, \mathfrak{Y}, \mathfrak{X})$ is a suitable system, and $\mathfrak{x} \in \mathfrak{P}$ if and only if $\theta^{-1}(\mathfrak{x})$ is non–degenerate, then it is easy to see that $\mathfrak{P}$ is a finite set

(= a set of finite cardinal number) and if $\Omega = \theta^{-1}(\mathfrak{P})$, then Ω is also finite.

The sets $\mathfrak{P}$ and Ω are called the exceptional sets of $(\theta, \mathfrak{Y}, \mathfrak{X})$. (The sets $\mathfrak{P}$ and Ω clearly depend upon the mapping θ; in particular, if $(\theta', \mathfrak{Y}', \mathfrak{X})$ is another suitable system with exceptional sets $\mathfrak{P}'$ and Ω', the sets $\mathfrak{P}$ and $\mathfrak{P}'$ need not be identical. In fact, it is possible to have two suitable systems $(\theta, \mathfrak{Y}, \mathfrak{X})$ and $(\theta', \mathfrak{Y}, \mathfrak{X})$ — note that the last two terms of the first triple are the same as the last two terms of the second — with different exceptional sets $\mathfrak{P}$ and $\mathfrak{P}'$ in $\mathfrak{X}$, and different exceptional sets Ω and Ω' in $\mathfrak{Y}$.) Notice that $\theta \,|\, (\mathfrak{Y} - \Omega)$ is a homeomorphism from $(\mathfrak{Y} - \Omega)$ onto $(\mathfrak{X} - \mathfrak{P})$, and further, that if $\mathfrak{y} \in \mathfrak{Y} - \Omega$, then $\mathfrak{y} = \theta^{-1}\theta\,(\mathfrak{y})$.

THEOREM: $\mathfrak{X}$ *is a mantoid if and only if there is a suitable system* $(\theta, \mathfrak{Y}, \mathfrak{X})$.

This theorem is proved by Roberts and Steenrod as the last of a sequence of theorems concerned with the problem of characterizing mantoids. Some of these theorems are to be employed in this paper and are catalogued here. Their work is based upon a lemma of fundamental importance.

6.2   LEMMA (Roberts and Steenrod):

Given:

1° *A closed 2—manifold,* X.

2° *A continuum,* K, *in* X.

3° *A component,* S, *of* $X - K$.

Conclusion: S *contains a 2—manifold,* M, *with boundary curves,* $\Gamma_1, \cdots, \Gamma_s$, *and* $S - M$ *has* s *components,* $H_1, \cdots, H_s$, *where:*

4° H_k *is an open cylinder,* $k = 1, \cdots, s$.

5° $\dot{H}_k$ *has two components, one is* Γ_k *and the other is a continuum,* Δ_k, *in* K, $k = 1, \cdots, s$.

Remark: If S contains a 2—manifold M^* with boundary curves $\Gamma_1^*, \cdots, \Gamma_r^*$, and $S - M^*$ has r components, $H_1^*, \cdots, H_r^*$, with properties 4° and 5° of the lemma, then $r = s$. Hence though the 2—manifold M is not uniquely determined in terms of S, the number s is. The component S will be said to have s *cylinders of approach to* K, and M will be called a 2—manifold *associated with* S. It is, of course, possible to find a sequence $\{M_n\}$ of 2-manifolds, M_n, $n = 1, 2, 3, \cdots$, associated with S such that $M_n \subset M_{n+1}^\circ$, $n = 1, 2, 3, \cdots$, and $\bigcup M_n = S$.

6.3   In the next few paragraphs it is necessary to concern oneself with $R(K)$, the 1-dimensional Betti number (taken mod 2) of a continuum K in a closed 2-manifold.

6.4   LEMMA:

Given:

1° *A closed 2-manifold, X.*

2° *A continuum, K, which is a proper subset of X.*

Conclusion: $R(K) = 0$ *if and only if each open set, G, contain-ing K contains a 2-cell, C, such that $K \subset C^\circ$. (If $R(K) = 0$ then $C^\circ - K$* is connected.)

6.5   THEOREM: *If $m: X \rightrightarrows \mathfrak{X}$ is a monotone mapping where X is a closed 2-manifold, and $\mathfrak{X}$ is non-degenerate, then X is homeomorphic to $\mathfrak{X}$ if and only if $R(m^{-1}(\mathfrak{x})) = 0$ for $\mathfrak{x} \in \mathfrak{X}$.*

6.6   THEOREM: *If $m: X \rightrightarrows \mathfrak{X}$ is a monotone mapping where X and $\mathfrak{X}$ are closed 2-manifolds, then $R(m^{-1}(\mathfrak{x})) = 0$ except for at most a finite number of points, $\mathfrak{x}$, in $\mathfrak{X}$.*

6.7   THEOREM: *If $m: X \rightrightarrows \mathfrak{X}$ is a monotone mapping where X is a closed orientable 2-manifold and $\mathfrak{X}$ is a generalized cactoid, then each true cyclic element of $\mathfrak{X}$ is an orientable 2-manifold.*

6.8   In proving the statement that if $\mathfrak{X}$ is a mantoid, then there is a suitable system $(\theta, \mathfrak{Y}, \mathfrak{X})$, the following result is obtained:

THEOREM:

Given: *A monotone mapping, $m: X \rightrightarrows \mathfrak{X}$, where X is a closed 2-mani-fold.*

Conclusion: *There exists*

1° *A closed 2-manifold, Y, which is orientable if X is orientable.*

2° *A suitable system, $(\theta, \mathfrak{Y}, \mathfrak{X})$, with exceptional sets $\mathfrak{P}$ in $\mathfrak{X}$ and Ω in $\mathfrak{Y}$* (see 6.1).

3° *A monotone mapping, $\mu: Y \rightrightarrows \mathfrak{Y}$, and a homeomorphism,*

$$h: [X - m^{-1}(\mathfrak{P})] \approx [Y - \mu^{-1}(\Omega)], \quad such\ that \quad m(x) = \theta \mu h(x) \quad if \quad x \in [X - m^{-1}(\mathfrak{P})].$$

6.9   In addition to these basic results of Roberts and Steenrod, the following theorem is found to be useful in the sequel. The proof is a simple exercise involving lemma 6.2.

THEOREM:

Given:

1° *A monotone mapping,* $\ m: X \rightrightarrows \mathfrak{X}, \ $ *where* $\ X \ $ *is a closed 2-manifold.*

2° *A region,* $\ \mathfrak{U}, \ $ *in* $\ \mathfrak{X}, \ $ *and a component,* $\ \mathfrak{R}, \ $ *of* $\ \dot{\mathfrak{U}}.$

3° *A sequence,* $\ \{\mathfrak{R}_n\}, \ $ *of regions,* $\ \mathfrak{R}_n, \ n = 1, 2, 3, \cdots,$ *such that:*

     (i) $\mathfrak{R}_n \subset \mathfrak{U}, \ n = 1, 2, 3, \cdots.$

     (ii) $\overline{\mathfrak{R}}_n \cap \dot{\mathfrak{U}} = \mathfrak{R}, \ n = 1, 2, 3, \cdots.$

     (iii) $\overline{(\mathfrak{U} - \mathfrak{R}_n)} \cap \mathfrak{R} = 0, \ n = 1, 2, 3, \cdots.$

     (iv) $Lim\ \overline{\mathfrak{R}}_n = \mathfrak{R}.$

4° $U = m^{-1}(\mathfrak{U}).$

Conclusion:

5° *There is a unique component,* $\ K, \ $ *of* $\ \dot{U} \ $ *in* $\ m^{-1}(\mathfrak{R}).$

6° $U \ $ *has a single cylinder of approach to* $\ K.$

7° $m(K) = \mathfrak{R}.$

6.10   The following theorem, though not proved by Roberts and Steenrod, is closely related to their work and may be considered an exercise in the techniques they employ.

THEOREM:

Given:

1° *A monotone mapping,* $\ m: X \rightrightarrows \mathfrak{X}, \ $ *where* $\ X \ $ *is a closed 2-manifold.*

2° *A free open arc,* $\ a, \ $ *of* $\ \mathfrak{X}.$

3° $\epsilon > 0.$

Conclusion: *There is a finite set of open arcs,* $\ a_1, \cdots, a_n,$ *such that:*

4° $\bar{a} = \bigcup \bar{a}_k.$

5° $\bar{a}_k \cap \bar{a}_{k+1} \ $ *is degenerate,* $\ k = 1, \cdots, (n - 1).$

6° $\quad a_k \cap a_j = 0, \quad k \neq j; \quad k, \; j = 1, \; \cdots, \; n.$

7° $\quad d(a_k) < \epsilon, \quad k = 1, \; \cdots, \; n.$

8° $\quad m^{-1}(a_k) \;$ *is an open cylinder.*

7. The Čech theory

The techniques here employed in algebraic topology are due principally to Eilenberg and Steenrod; this is true, in particular, in reference to what they have picturesquely described as "diagram chasing." *The coefficient group is taken as the additive group of integers.*

<u>7.1</u>   Though the Čech cohomology groups can be defined for general topological spaces, it appears both convenient and compatible with the general tenor of this paper to restrict the discussion to compact (= bicompact) spaces.

It is recalled that a basic notion is that of a pair, (X, A), consisting of a compact space, X, and a closed subset, A. In the event A is empty, the pair (X, A) will sometimes be written (X).

A mapping $f: (X, A) \to (Y, B)$ is a mapping $f: X \to Y$ such that $f(A) \subset B$. The most common mappings considered are *identity mappings*, and they occur automatically if $(X, A) \subset (Y, B)$, that is, $X \subset Y$ and $A \subset B$. Under these circumstances the mapping $i(x) = x, \; x \in X$ provides the identity mapping $i: (X, A) \to (Y, B)$. The notation $f: (X, A) \rightrightarrows (Y, B)$ is sometimes used to indicate that $f(X) = Y$ and $f(A) = B$, while $f: (X, A) \approx (Y, B)$ means that $f: (X, A) \rightrightarrows (Y, B)$ and f is a homeomorphism.

The Čech cohomology theory associates an Abelian group, $H^q(X, A)$, with each pair, (X, A), for every integer, $q \geq 0$. The group is known as the q-dimensional relative cohomology group of $X \bmod A$ (with integral coefficients), and the topology on it is considered *discrete*.

For each pair, (X, A), and every $q \geq 0$ there is a homomorphism,
$$\delta_q : H^q(A) \to H^{q+1}(X, A),$$
known as the *coboundary operator*. In addition, if $f: (X, A) \to (Y, B)$ is a mapping, then for each $q \geq 0$ there is a homomorphism from $H^q(Y, B)$ into $H^q(X, A)$, said to be *induced* by f and designated by
$$f_q^* : H^q(Y, B) \to H^q(X, A).$$

(It is to be noted that the induced homomorphism runs in a direction opposite to the original mapping.) In the sequel the subscript q is dropped since no confusion can arise in replacing δ_q and f_q^* by δ and f^*, respectively, as long as the groups are indicated. The notation $f^*:H^q(Y,\ B) \approx H^q(X,\ A)$ will indicate that f^* is an isomorphism from $H^q(Y,\ B)$ *onto* $H^q(X,\ A)$. Such an isomorphism is to be known as an *isomorphism onto*, the two words being used as a single noun.

If $Y' \subset Y$, $B' \subset B$, $f(X) \subset Y'$ and $f(A) \subset B'$, then $f:(X,\ A) \rightarrow (Y,\ B)$ is also a mapping from the pair $(X,\ A)$ into the pair $(Y',\ B')$ and induces a homomorphism from $H^q(Y',\ B')$ into $H^q(X,\ A)$ for each $q \geqq 0$. Under these conditions the notation $\bar{f}^*$, or f^* with some additional symbol, will be used for this latter homomorphism to avoid confusion with $f^*:H^q(Y,\ B) \rightarrow H^q(X,\ A)$. In the event $(X,\ A) \subset (Y,\ B)$, then a notation such as $i^*:H^q(Y,\ B) \rightarrow H^q(X,\ A)$ will be used to indicate the homomorphism induced by the identity mapping, i, without making any initial statement that i is the identity; nor will the letter i alone be used for this purpose. For example, if $(X,\ A)$ is a pair, then the notation $a^*:H^q(X,\ A) \rightarrow H^q(X)$ means that a^* is the homomorphism induced by the mapping $a(x) = x$, $x \in X$.

It will be helpful to recall a few theorems, it being understood that if no comment is made concerning the dimension q, then the theorem is true for $q \geqq 0$.

7.2   THEOREM: *If $f:(X,\ A) \approx (Y,\ B)$, then $f^*:H^q(Y,\ B) \approx H^q(X,\ A)$ and* $(f^*)^{-1} = (f^{-1})^*$.

7.3   THEOREM: *If $f:(X,\ A) \rightarrow (Y,\ B)$ and $g:(Y,\ B) \rightarrow (Z,\ C)$ are mappings, while $h:(X,\ A) \rightarrow (Z,\ C)$ is the composition (gf), then one has commutativity in the following diagram:*

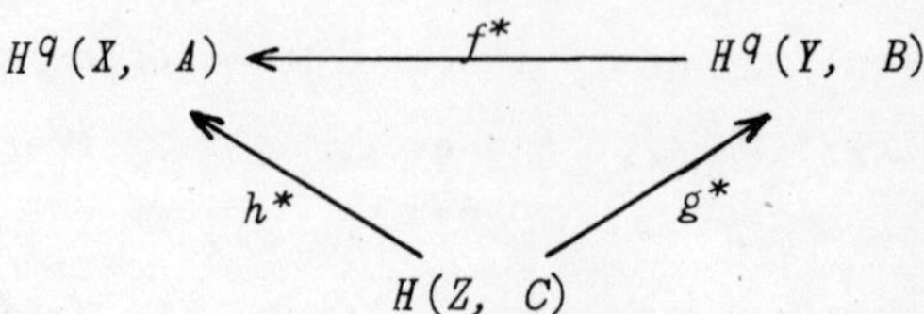

That is, if $z \in H^q(Z,\ C)$, then $f^*g^*(z) = h^*(z)$ or simply $f^*g^* = h^* = (gf)^*$.

7.4   THEOREM: *If $f:(X,\ A) \rightarrow (Y,\ B)$ is a mapping, and $\underline{f} = f|A$, then one has commutativity in the diagram below:*

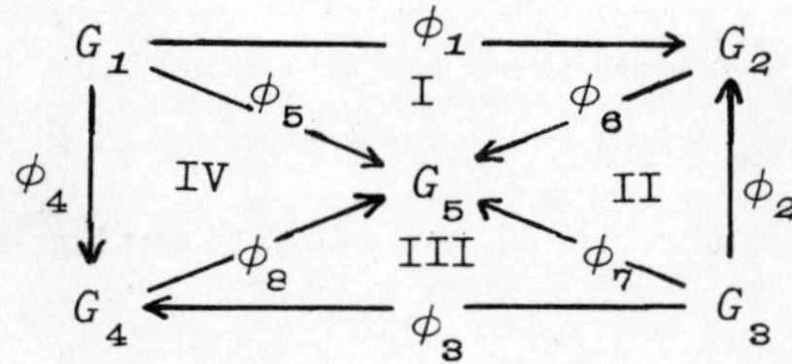

$$H^{q+1}(X,\ A) \xleftarrow{\quad f^* \quad} H^{q+1}(Y,\ B)$$

That is, $f^*\delta = \delta f^*$.

7.5   Explicit references to 7.2–7.4 will be given rather rarely in the
sequel. In view of this fact and the continued use of the word *commutative*,
it should be stated that the proofs of many of the theorems to follow will
involve the use of diagrams. For example, a diagram of the following character may appear

where G_i is a group and ϕ_j a homomorphism, $i = 1,\ \cdots,\ 5;\ j = 1,\ \cdots,\ 8.$
The diagram consists of four subdiagrams labeled I, II, III and IV. Such subdiagrams within a diagram will be called *boxes*. Relative to each box it will
be possible to ask whether or not commutativity holds; for example, in box II
it is possible to go from G_3 to G_5 by ϕ_7 or by $\phi_6\phi_2$.

        If the statement *commutativity holds in every box* is found in the
proof of a theorem, then it will follow — unless some other reason is given
explicitly — from 7.3 and 7.4.

7.6   THEOREM: *If $f:(X,\ A) \to (Y,\ B)$ is a mapping such that f maps
$(X - A)$ homeomorphically onto $(Y - B)$ then $f^*:H_q(Y,\ B) \approx H^q(X,\ A)$.*

7.7   THEOREM: *If U is an open subset of X and is contained in a
closed subset A, then $i^*:H^q(X,\ A) \approx H^q(X - U,\ A - U)$.*

        Remark: This is the strong form for the *excision axiom* of Eilenberg
and Steenrod. In referring to this result later, the phrase "by excision"
will frequently be employed.

7.8   THEOREM: *If A is a finite set of points of X, then $i^*:H^q(X,\ A)$
$\approx H^q(X)$ for $q > 1$.*

7.9   THEOREM: *If $f_n:(X,\ A) \to (Y,\ B)$ is a mapping, $n = 0,\ 1,\ 2,\ \cdots,$
and $f_n \rightrightarrows f_0,$ then $f_n^* \to f_0^*;$ that is, if $\mathbf{y} \in H^q(Y,\ B),$ then there is an
n_0 such that for $n > n_0$ it is true that $f_n^*(\mathbf{y}) = f_0^*(\mathbf{y}).$* (Recall that, as
a space, $H^q(X,\ A)$ is discrete.)

8.   The locally compact theory

The locally compact theory (H. Cartan [2]) is employed here simply
because of the considerable abbreviation it permits both in the statement and
also in the proof of almost every theorem to follow in the paper.   The ordinary
Čech theory, however, is completely adequate, albeit cumbersome, in this en—
deavor.

8.1   Suppose X is a locally compact space and A is a closed subset of
X.   Let $(X - A) \equiv Y$ be compactified by the addition of a set $\Re$;   that is,
there is a compact space, $\mathfrak{Y}$,   such that $Y \subset \mathfrak{Y},\ \mathfrak{Y} - Y = \Re$ and $\Re$ is *closed*.
Suppose further that $\check{Y}$ is the usual compactification of Y by the addition
of a single point $\check{y}$.

There is a mapping $\nu : \mathfrak{Y} \to \check{Y}$,   uniquely determined by the requirement
that ν is the identity on Y.   In fact,

$$\nu(\mathfrak{y}) = \begin{cases} \mathfrak{y} & \text{if } \mathfrak{y} \in Y \\ \check{y} & \text{if } \mathfrak{y} \in \Re. \end{cases}$$

The mapping is onto $\check{Y}$ unless $\Re = 0$.   In view of 7.6, $\nu : (\mathfrak{Y},\ \Re) \to (\check{Y},\ \check{y})$
induces an isomorphism:

$$\nu : H^q(\check{Y},\ \check{y}) \approx H^q(\mathfrak{Y},\ \Re)$$

called the natural isomorphism from $H^q(\check{Y},\ \check{y})$ onto $H^q(\mathfrak{Y},\ \Re)$.

If $\mathfrak{Y}_1$ and $\mathfrak{Y}_2$ are two compactifications of Y by the addition of
sets $\Re_1$ and $\Re_2$, respectively, then, following the notation introduced
above, one has:

$$\nu_i : H^q(\check{Y},\ \check{y}) \approx H^q(\mathfrak{Y}_i,\ \Re_i),\ i = 1,\ 2.$$

If $\phi_2^1 = \nu_2(\nu_1)^{-1},$ then

$$\phi_2^1 : H^q(\mathfrak{Y}_1,\ \Re_1) \approx H^q(\mathfrak{Y}_2,\ \Re_2).$$

In other words, beginning with a pair $(X,\ A)$, where X is a
locally compact space and A is a closed subset, it is possible to arrive

at a definite collection, $\{H^q(\mathfrak{Y}, \mathfrak{R})\}$, of groups, $H^q(\mathfrak{Y}, \mathfrak{R})$, such that each ordered pair $[H^q(\mathfrak{Y}_1, \mathfrak{R}_1), H^q(\mathfrak{Y}_2, \mathfrak{R}_2)]$ gives rise to a particular isomorphism, $\phi_2^1 : H^q(\mathfrak{Y}_1, \mathfrak{R}_1) \approx H^q(\mathfrak{Y}_2, \mathfrak{R}_2)$, called the natural isomorphism from $H^q(\mathfrak{Y}_1, \mathfrak{R}_1)$ onto $H^q(\mathfrak{Y}_2, \mathfrak{R}_2)$. Moreover, ϕ_1^1 is the identity, $\phi_2^1 = (\phi_1^2)^{-1}$, and for any three groups $H^q(Y_i, K_i)$, $i = 1, 2, 3$, it is true that $\phi_3^1 = \phi_3^2 \phi_2^1$. Under these circumstances, $\{H^q(\mathfrak{Y}, \mathfrak{R})\}$ may be regarded, under the identification provided by the ϕ's, as a single group to be designated by $H^q(X, A)$. Notice that, under this convention, $H^q(X, A)$ and $H^q(X - A)$ — that is, $H^q(X - A, 0)$ — are simply two ways of writing the same group. Hence the locally compact theory is really an absolute rather than a relative theory. It is convenient, however, to use the notation $H^q(X, A)$ since, if X is compact, then $H^q(X, A)$ may be regarded as the Čech cohomology group since X is a compactification of $X - A$ by the addition of A.

8.2   Suppose X and Y are locally compact spaces while $\check{X}$ and $\check{Y}$ are compactifications by the addition of points $\check{x}$ and $\check{y}$, respectively. Given a mapping $f : X \to Y$, define $\check{f}(x) = f(x)$, $x \in X$, and $\check{f}(\check{x}) = \check{y}$. It is well known that $\check{f}$ is a mapping if and only if f is a *compact mapping*; that is, C compact implies $f^{-1}(C)$ compact. Assuming f to be compact, $\check{f} : (\check{X}, \check{x}) \to (\check{Y}, \check{y})$ induces a homomorphism $\check{f}^* : H^q(\check{Y}, \check{y}) \to H^q(\check{X}, \check{x})$ in the Čech theory. If $\mathbf{y} \in H^q(Y)$, then it may be regarded as an element of the group $H^q(\check{Y}, \check{y})$ and there is a unique element $\mathbf{x} \in H^q(X)$ which may be regarded as $\check{f}^*(\mathbf{y})$. On defining $f^*(\mathbf{y}) = \mathbf{x}$ one obtains a homomorphism, $f^* : H^q(Y) \to H^q(X)$, said to be induced by the compact mapping $f : X \to Y$. The homomorphism f^* has been defined in terms of particular compactifications of X and Y. However, if $\mathfrak{X}$ and $\mathfrak{Y}$ are compactifications of X and Y by the addition of sets $\mathfrak{R}$ and $\mathfrak{J}$, respectively, and if $f : X \to Y$ can be extended to a mapping $\bar{f} : (\mathfrak{X}, \mathfrak{R}) \to (\mathfrak{Y}, \mathfrak{J})$, then it follows, under the principle of identification, that $\bar{f}^* = \check{f}^* = f^*$.

8.3   The principal use to be made of the locally compact theory is to replace, on occasion, the relative Čech theory by the locally compact theory. To illustrate, suppose X is a compact space and G an open subset of X. Then G is locally compact and $H^q(X, X - G)$ or $H^q(\bar{G}, \dot{G})$ in the Čech theory is simply $H^q(G)$ in the locally compact.

Homomorphisms in the locally compact theory will usually arise in the following manner. If $f:X \rightrightarrows \mathcal{X}$ is a mapping where X and $\mathcal{X}$ are compact spaces, while $\mathcal{G}$ is an open set in $\mathcal{X}$ and $G = f^{-1}(\mathcal{G})$, consider $g = f|G$. The mapping $g:G \rightrightarrows \mathcal{G}$ is compact, and there are several compactifications of the locally compact spaces G and $\mathcal{G}$ such that g can be extended to a mapping from the compactification of G onto the compactification of $\mathcal{G}$. For example, X is a compactification of G while $\mathcal{X}$ is a compactification of $\mathcal{G}$ and $f:X \rightrightarrows \mathcal{X}$ is an extension of g. Further, $\overline{G}$ is a compactification of G while $\overline{\mathcal{G}}$ is a compactification of $\mathcal{G}$, and if $\overline{f} = f|\overline{G}$ then $\overline{f}:\overline{G} \rightrightarrows \overline{\mathcal{G}}$ is an extension of g. Hence $g^*:H^q(\mathcal{G}) \to H^q(G)$ may be regarded as $f^*:H^q(\mathcal{X}, \mathcal{X} - \mathcal{G}) \to H^q(X, X - G)$ or $\overline{f}^*:H^q(\overline{\mathcal{G}}, \dot{\mathcal{G}}) \to H^q(\overline{G}, \dot{G})$; sometimes one interpretation will be convenient, at other times another. This freedom of choice will be a great help later.

<u>8.4</u>   In connection with induced homomorphisms and relative to the remarks of the preceding paragraphs, it is well to consider an anomalous situation which obtains in the event Y is a locally compact space and X is an open subset of Y. Then X is a locally compact space, but the identity mapping $i:X \to Y$ will not always be a compact mapping. Consequently, $i^*:H^q(Y) \to H^q(X)$ will not always have meaning. However, there is *always* a naturally determined homomorphism from $H^q(X)$ into $H^q(Y)$ — note that the homomorphism is *not* from $H^q(Y)$ into $H^q(X)$, which is the direction in which an induced homomorphism would run.

The homomorphism is defined in the following manner. Let $\mathcal{Y}$ be a compactification of Y by the addition of a set $\mathcal{R}$. Now $\mathcal{Y}$ is also a compactification of X by the addition of the set $(\mathcal{Y} - X)$. Moreover, $\mathcal{R} \subset (\mathcal{Y} - X)$, hence the identity mapping $j:(\mathcal{Y}, \mathcal{R}) \to (\mathcal{Y}, \mathcal{Y} - X)$ induces a homomorphism

$$j^*:H^q(\mathcal{Y}, \mathcal{Y} - X) \to H^q(\mathcal{Y}, \mathcal{R}).$$

However, j^* may be regarded as a homomorphism

$$j:H^q(X) \to H^q(Y).$$

(It is easy to see that j is independent of the particular compactification of Y employed.) This is the homomorphism mentioned in the preceding paragraph, and as a notational convention *such homomorphisms will always be indicated by the use of a German letter without an asterisk.*

It should be noted that if $i:X \to Y$ is a compact mapping, then $i^* = j^{-1}$.

8.5   It is obvious that theorems in the locally compact theory will be proved by going to the Čech theory. Some of the results to be employed in the sequel are tabulated here.

THEOREM: *If $f:X \rightrightarrows Y$ is a homeomorphism, where X and Y are locally compact spaces, then $f^*:H^q(Y) \approx H^q(X)$ and $(f^*)^{-1} = (f^{-1})^*$.* (Cf. 7.2.)

8.6   THEOREM: *If G is an open subset of a locally compact space X, and $(X - G)$ is a finite set of points, then for $q > 1$, $i:H^q(G) \approx H^q(X)$.* (Cf. 7.8.)

8.7   THEOREM:

Given:

1°. *A compact mapping, $f:X \to \mathfrak{X}$, where X and $\mathfrak{X}$ are locally compact spaces.*

2°. *$\mathfrak{G}$ is an open subset of $\mathfrak{X}$, and $G = f^{-1}(\mathfrak{G})$.*

3°. *$\bar{f} = f|G$.*

Conclusion: *Commutativity holds in the following diagram:*

$$
\begin{array}{ccc}
H^q(X) & \xleftarrow{\quad i \quad} & H^q(G) \\
f^* \big\uparrow & & \big\uparrow \bar{f}^* \\
H^q(\mathfrak{X}) & \xleftarrow{\quad j \quad} & H^q(\mathfrak{G})
\end{array}
$$

As an example of a proof, note that commutativity holds in the following diagram of *mappings* where $\underline{\check{f}} = \check{f}$:

$$
\begin{array}{ccc}
(\check{X},\ \check{x}) & \xrightarrow{\quad i \quad} & (\check{X},\ \check{X} - G) \\
\check{f} \big\downarrow & & \big\downarrow \check{\bar{f}} \\
(\check{\mathfrak{X}},\ \check{\mathfrak{x}}) & \xrightarrow{\quad j \quad} & (\check{\mathfrak{X}},\ \check{\mathfrak{X}} - \mathfrak{G})
\end{array}
$$

Hence, by 7.3, it holds for the following diagram of *induced homomorphisms:*

$$
\begin{array}{ccc}
H^q(\check{X},\ \check{x}) & \xleftarrow{\quad i^* \quad} & H^q(\check{X},\ \check{X} - G) \\
\check{f}^* \big\uparrow & & \big\uparrow \check{\bar{f}}^* \\
H^q(\check{\mathfrak{X}},\ \check{\mathfrak{x}}) & \xleftarrow{\quad j^* \quad} & H^q(\check{\mathfrak{X}},\ \check{\mathfrak{X}} - \mathfrak{G})
\end{array}
$$

But this is simply another way of writing the conclusion of the theorem.

Remark: A situation which will often occur in the sequel is one in which $f: X \rightrightarrows \mathfrak{X}$ will be a mapping where X and $\mathfrak{X}$ are compacta, $\mathfrak{U}$ and $\mathfrak{V}$ will be open subsets of $\mathfrak{X}$ with $U = f^{-1}(\mathfrak{U})$ and $V = f^{-1}(\mathfrak{V})$, while $\mathfrak{U} \supset \mathfrak{V}$. If $\underline{f} = f|U$ and $\bar{f} = f|V$, then $\underline{f}$ and $\bar{f}$ are compact mappings, and the theorem guarantees that commutativity holds in the following diagrams:

$$
\begin{array}{ccc}
H^q(U) & \xleftarrow{\quad a \quad} & H^q(V) \\
\underline{f}^* \big\uparrow & & \big\uparrow \bar{f}^* \\
H^q(\mathfrak{U}) & \xleftarrow{\quad b \quad} & H^q(\mathfrak{V})
\end{array}
$$

$$
\begin{array}{ccc}
H^q(X) & \xleftarrow{\quad c \quad} & H^q(U) \\
f^* \big\uparrow & & \big\uparrow \underline{f}^* \\
H^q(\mathfrak{X}) & \xleftarrow{\quad b \quad} & H^q(\mathfrak{U})
\end{array}
$$

9. Cohomology theory of 2-manifolds and canonical regions

It is well known that if X is a 2-manifold, then $H^2(X, \dot{X})$ is cyclic infinite or of order 2, where $\dot{X} = 0$ if X is closed. In the first case the 2-manifold is said to be *orientable,* in the second *non-orientable.*

The following theorems are obtained by re-wording the classical theorems on 2-manifolds, and are recorded here for future convenience.

9.1   THEOREM: *If U is a canonical region on a 2-manifold X, then $H^2(U)$ is cyclic infinite or of order 2.*

Definition: A canonical region, U, is said to be *orientable* [*non-orientable*] if and only if $H^2(U)$ is cyclic infinite [of order 2].

Remark: If M is a 2-manifold associated with a canonical region U (see 5.3), then clearly U is orientable if and only if M is orientable.

9.2   THEOREM:

      Given:

           1°. *A closed orientable 2-manifold, X.*

           2°. *A canonical region, U, on X.*

      Conclusion: $i^*: H^2(X, X - U) \approx H^2(X)$; *that is,* $i: H^2(U) \approx H^2(X)$, *and hence U is orientable.*

9.3   THEOREM:

Given:

    1° *Two canonical regions, U and V,   on a 2-manifold.*

    2° $U \supset V$ *and* U *is orientable.*

Conclusion: $\mathfrak{k}: H^2(V) \approx H^2(U)$

Immediate consequence: V *is orientable.*

9.4   THEOREM:

Given:

    1° *Three canonical regions, U_1, U_2 and U ,   on a 2-manifold.*

    2° $U_1 \supset U_2 \supset U_3$.

    3° U_1 *is orientable.*

Conclusion: *In the diagram below all the homomorphisms are iso-morphisms onto, and commutativity holds:*

$$
\begin{array}{ccc}
H^2(U_1) & \xleftarrow{\quad i \quad} & H^2(U_2) \\
 & \nwarrow{\scriptstyle j} \qquad \nearrow{\scriptstyle \mathfrak{k}} & \\
 & H^2(U_3) &
\end{array}
$$

As an example of a proof, notice that the homomorphisms are iso-morphisms onto, in virtue of 9.3, and the diagram of the theorem may be regarded as:

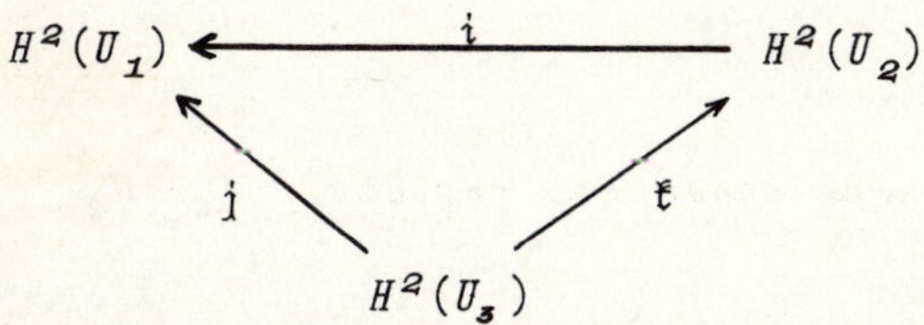

$$
\begin{array}{ccc}
H^2(\bar{U}_1,\ \dot{U}_1) & \xleftarrow{\quad i^* \quad} & H^2(\bar{U}_1,\ \bar{U}_1 - U_2) \\
 & \nwarrow{\scriptstyle j^*} \qquad \nearrow{\scriptstyle k^*} & \\
 & H^2(\bar{U}_1,\ \bar{U}_1 - U_3) &
\end{array}
$$

Remark: If the Čech theory alone were to be used, the theorem, to be useful in the sequel, would have for its diagram:

$$
\begin{array}{ccccc}
H^2(\bar{U}_1,\ \dot{U}_1) & \xleftarrow{\ i^* \ } & H^2(\bar{U}_1,\ \bar{U}_1 - U_2) & \xrightarrow{\ a^* \ } & H^2(\bar{U}_2,\ \dot{U}_2) \\
{\scriptstyle j^*}\uparrow & & & & \uparrow{\scriptstyle b^*} \\
H^2(\bar{U}_1,\ \bar{U}_1 - U_3) & \xrightarrow{\ k^* \ } & H^2(\bar{U}_3,\ \dot{U}_3) & \xleftarrow{\ c^* \ } & H^2(\bar{U}_2,\ \bar{U}_2 - U_3)
\end{array}
$$

This single example should be a convincing demonstration of the advantages of using the locally compact theory.

9.5   THEOREM:

Given:

1° *Two canonical regions, U_1 and U_2, on a closed 2-manifold.*

2° $U_1 \supset U_2$.

3° G_1 and G_2 *are open 2-cells in U_2.*

4° U_1 *is orientable.*

Conclusion: *In the diagram below all the homomorphisms are isomorphisms onto, and commutativity holds:*

$$
\begin{array}{ccc}
H^2(G_1) & \xrightarrow{\ \ c\ \ } & H^2(U_1) \\
\downarrow{\scriptstyle a} & & \uparrow{\scriptstyle b} \\
H^2(U_2) & \xleftarrow{\ \ b\ \ } & H^2(G_2)
\end{array}
$$

9.6   THEOREM:

Given:

1° *Three canonical regions, U_1, U_2 and U, on a closed 2-manifold.*

2° $U_i \subset U$, $i = 1, 2$.

3° G_1 and G_2 *are open 2-cells in U_i, $i = 1, 2$.*

4° U *is orientable.*

Conclusion: *In the diagram of the preceding theorem all the homomorphisms are isomorphisms onto, and commutativity holds.*

9.7   If G_1, G_2 and G_3 are groups while $\phi_1 : G_1 \to G_2$, $\phi_2 : G_2 \to G_3$ and $\phi_3 : G_1 \to G_3$ are homomorphisms, then the statement *negative commutativity* holds in the diagram below

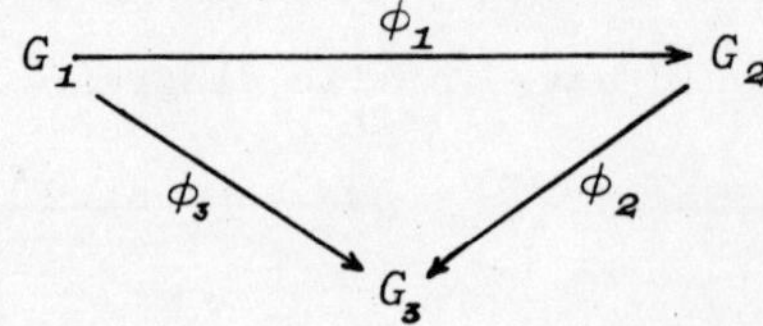

means that

$$\phi_3(g_1) = -\,\phi_2\phi_1(g_1), \quad g_1 \in G_1$$

or simply,

$$\phi_3 = -\,\phi_2\phi_1.$$

One can now state the following result in which the hypothesis can be considerably weakened; the statement here is tailored to the contemplated applications.

THEOREM:

Given:

1°. *Three canonical regions, U, V and W, on a closed 2-manifold.*

2°. $W = U \cup V$.

3°. *U is orientable and V is an open cylinder.*

4°. *$U \cap V$ has two components, each an open cylinder.*

5°. *G_1 and G_2 are open 2-cells in different components of $U \cap V$.*

Conclusion: *All the homomorphisms in the diagram below are isomorphisms onto, and if W is orientable [non-orientable] one has commutativity [negative commutativity]:*

$$
\begin{array}{ccc}
H^2(G_1) & \xrightarrow{\ \ c\ \ } & H^2(V) \\
{\scriptstyle a}\big\downarrow & & \big\uparrow{\scriptstyle b} \\
H^2(U) & \xleftarrow{\ \ b\ \ } & H^2(G_2)
\end{array}
$$

9.8   It is important to have some theorems concerning the behavior of the coboundary homomorphism; for example,

THEOREM: *If M is an orientable 2-manifold and $\dot{M}$ consists of a single 1-sphere, then $\delta : H^1(\dot{M}) \approx H^2(M, \dot{M})$.*

If M is an orientable 2-manifold with boundary curves $J_1, \cdots, J_n$, $n > 1$, let $a_k : J_k \to \dot{M}$ be the identity mapping and $a_k^* : H^1(\dot{M}) \to H^1(J_k)$ the induced homomorphism; suppose $V_k = $ kernel a_k^* and $W_k = (V_1 \cap \cdots \cap \hat{V}_k \cap \cdots \cap V_n)$, $k = 1, \cdots, n$, where the notation $\hat{V}_k$ means that V_k is omitted. Let

$$\bar{a}_k^* = a_k^* | W_k, \quad k = 1, \cdots, n$$

and

$$\delta_k = \delta | W_k, \quad k = 1, \cdots, n.$$

It can be shown that:

$$H^1(\dot{M}) = W_1 \oplus \cdots \oplus W_n,$$

$$\bar{a}_k^* : W_k \approx H^1(J_k), \quad k = 1, \cdots, n,$$

$$\delta_k : W_k \approx H^2(M, \dot{M}).$$

Hence, if

$$\delta_k^* = \delta_k(\bar{a}_k^*)^{-1}, \quad k = 1, \cdots, n,$$

one has the following generalization of the above result.

THEOREM: $\delta_k^* : H^1(J_k) \approx H^2(M, \dot{M})$, $k = 1, \cdots, n$.

The notation $\delta_k^* : H^1(J_k) \approx H^2(M, \dot{M})$ *will often be abbreviated to read* $\delta^* : H^1(J_k) \approx H^2(M, \dot{M})$, the presence of the groups making it impossible for confusion to arise.

If $\boldsymbol{j}_k$ is a generator for $H^1(J_k)$, then $\delta_k^*(\boldsymbol{j}_k)$ is a generator, $\boldsymbol{m}_k$, for $H^2(M, \dot{M})$, said to be *determined by* $\boldsymbol{j}_k$, $k = 1, \cdots, n$. In other words, an orientation of J_k thus induces an orientation of M, $k = 1, \cdots, n$. Conversely, if $\boldsymbol{m}$ is a generator for $H^2(M, \dot{M})$, then $(\delta_k^*)^{-1}(\boldsymbol{m})$ is a generator, $\boldsymbol{j}_k$, for $H^1(J_k)$, said to be *determined by* $\boldsymbol{m}$, $k = 1, \cdots, n$. In other words, an orientation of M thus induces an orientation of J_k, $k = 1, \cdots, n$.

The set $\dot{M}$ is said to be *oriented* if and only if an orientation has been selected for each J_k, $k = 1, \cdots, n$. It has been shown that an orientation of M induces an orientation of J_k, $k = 1, \cdots, n$, and hence an orientation of $\dot{M}$. An orientation of $\dot{M}$ is said to be *concordant* if and only if it is induced by an orientation of M. (Consequently, there are exactly two concordant orientations of $\dot{M}$.) It is easy to see that an orientation of $\dot{M}$ is concordant if and only if the orientations of M induced by the selected orientation of J_k is independent of $k = 1, \cdots, n$.

Now suppose that M and $\mathfrak{M}$ are homeomorphic orientable 2-manifolds, while $\dot{h}$ is a homeomorphism from $\dot{M}$ onto $\dot{\mathfrak{M}}$. If $J_1, \cdots, J_n$ $[\mathfrak{J}_1, \cdots, \mathfrak{J}_n]$ are the boundary curves of $\dot{M}$ $[\dot{\mathfrak{M}}]$, then it may be assumed that $\dot{h}(J_k) = \mathfrak{J}_k$, $k = 1, \cdots, n$. If $h_k = \dot{h}|J_k$, then h_k is a homeomorphism from J_k onto $\mathfrak{J}_k$ and $h_k^* : H^1(\mathfrak{J}_k) \approx H^1(J_k)$. If $\dot{M}$ is oriented, then a generator, $\boldsymbol{j}_k$, has been selected for $H^1(J_k)$, $k = 1, \cdots, n$. If $\boldsymbol{i}_k = (h_k^*)^{-1}(\boldsymbol{j}_k)$, then $\boldsymbol{i}_k$ is a generator for $H^1(\mathfrak{J}_k)$, $k = 1, \cdots, n$. The generators $\boldsymbol{i}_i, \cdots, \boldsymbol{i}_k$ provide an orientation of $\dot{\mathfrak{M}}$ said to be *induced by the homeomorphism* $\dot{h}$ *and the orientation of* $\dot{M}$. In particular, the homeomorphism $\dot{h}$ is said to *carry a concordant orientation of* $\dot{M}$ *into a concordant orientation of* $\dot{\mathfrak{M}}$ if and only if given a concordant orientation of $\dot{M}$, the orientation induced on $\mathfrak{M}$

by $\dot{h}$ and the given orientation of $\dot{M}$ is a concordant orientation of $\dot{\mathfrak{M}}$.

These definitions will be employed in the statement of a theorem which, together with other key theorems of this paper, is recorded in the last section of this chapter.

9.9   Using the notations leading up to the second theorem of 9.8, fix $k = 1, \cdots, n$ and suppose $|\mathfrak{M}_k|$ is the upper semi-continuous decomposition of M whose sets are $J_1, \cdots, \hat{J}_k, \cdots, J_n$ and single points of $M^o \cup J_k$. Let $\mu_k : M \rightrightarrows \mathfrak{M}_k$ be the associated mapping, $\mu_k(J_k) = \mathfrak{I}_k$, and $\mu_k | J_k = \underline{\mu}_k$. (See Youngs [12].)

The space $\mathfrak{M}_k$ is clearly a 2-manifold with boundary $\mathfrak{I}_k$. Hence $\delta : H^1(\mathfrak{I}_k) \approx H^2(\mathfrak{M}_k, \dot{\mathfrak{M}}_k)$. One now has the following result.

THEOREM: *In the diagram below all the homomorphisms are isomorphisms onto, and commutativity holds:*

$$
\begin{array}{ccc}
H^2(M,\ \dot{M}) & \xleftarrow{\quad\delta^*\quad} & H^1(J_k) \\[2pt]
\mu_k^* \uparrow & & \uparrow \underline{\mu}_k^* \\[2pt]
H^2(\mathfrak{M}_k,\ \mu_k(\dot{M})) \xrightarrow{\ a^*\ } H^2(\mathfrak{M}_k,\ \dot{\mathfrak{M}}_k) & \xleftarrow{\ \delta\ } & H^1(\mathfrak{I}_k)
\end{array}
$$

10.   Key theorems

The object of this section, as the title implies, is to catalogue some theorems which are at the heart of the solution of the representation problem.

10.1   THEOREM (Radó [8, p.425]):   *If $\{f_n\}$ is a sequence of mappings, $f_n : X \rightrightarrows \mathfrak{X}$, $n = 1, 2, 3, \cdots$, where X is Peanian, and $l : \mathfrak{X} \rightarrow Y$ is a light mapping such that $\{lf_n\}$ converges uniformly, then there is a subsequence, $\{f_{n_k}\}$, of $\{f_n\}$ which is uniformly convergent.*

10.2   THEOREM (Whyburn [10, p.174]):   *If $\{m_n\}$ is a sequence of monotone mappings, $m_n : X \rightrightarrows \mathfrak{X}$, $n = 1, 2, 3, \cdots$, where X is Peanian, and $m_n \rightrightarrows m_o$, then the mapping $m_o : X \rightrightarrows \mathfrak{X}$ is monotone.*

10.3   MODIFICATION THEOREM (Youngs [12]):   *If $m : X \rightrightarrows \mathfrak{X}$ is a monotone mapping, where X is a compactum, $\mathfrak{X}$ is a 2-cell, and $m^{-1}(\mathfrak{X}^o)$ is an open 2-cell, then there is a monotone mapping $\mu : X \rightrightarrows \mathfrak{X}$ such that:*

$1^o.\quad \mu(x) = m(x)\quad if\quad x \in m^{-1}(\dot{\mathfrak{X}}).$

$2°$ $x = \mu^{-1}\mu(x)$ if $x \not\in m^{-1}(\dot{\mathfrak{X}})$.

Some consequences of the modification theorem follow.

10.4   THEOREM: *If* $m:X \rightrightarrows \mathfrak{X}$ *is a monotone mapping where* X *and* $\mathfrak{X}$ *are closed 2-manifolds, then there is a 2-cell,* $\mathfrak{C}$, *on* $\mathfrak{X}$ *and a monotone mapping* $\mu:X \rightrightarrows \mathfrak{X}$ *such that:*

$1°$ $\mu^{-1}(\mathfrak{C}^O) = m^{-1}(\mathfrak{C}^O)$.

$2°$ *If* $x \not\in \mu^{-1}(\mathfrak{C}^O)$,   *then* $\mu(x) = m(x)$.

$3°$ *If* $x \in \mu^{-1}(\mathfrak{C}^O)$,   *then* $x = \mu^{-1}\mu(x)$.

Proof: It follows from theorem 6.6 that there is a 2-cell $\mathfrak{C}$ such that $m^{-1}(\mathfrak{C}^O)$ is an open 2-cell. Let $Y = m^{-1}(\mathfrak{C})$, $G = m^{-1}(\mathfrak{C}^O)$ and $\bar{m} = m|Y$. Then $\bar{m}:Y \rightrightarrows \mathfrak{C}$ fulfills the requirements of the modification theorem. Consequently, there is a mapping $\bar{\mu}:Y \rightrightarrows \mathfrak{C}$ which agrees with $\bar{m}$ on $(Y - G)$ and has the property that $x \in G$ implies $x = \mu^{-1}\mu(x)$. The mapping μ of the theorem is simply defined to be m on $(X - G)$ and $\bar{\mu}$ on G.

10.5   APPROXIMATION THEOREM (Youngs [12]): *If* $m:X \rightrightarrows \mathfrak{X}$ *is a monotone mapping where* X *and* $\mathfrak{X}$ *are homeomorphic 2-manifolds, and* $\epsilon > 0$, *then there is a homeomorphism* $h_\epsilon:X \rightrightarrows \mathfrak{X}$ *such that* $\bar{\rho}\{m, h_\epsilon\} < \epsilon$.

Remark: The proof of the theorem shows that if $\mathfrak{C}_1$, $\cdots$, $\mathfrak{C}_n$ are 2-cells on $\mathfrak{X}$ which are disjoint in pairs, and $m^{-1}(\mathfrak{x})$ is degenerate for $\mathfrak{x} \in \bigcup \mathfrak{C}_k$, then h_ϵ may be determined so that $x \in \bigcup m^{-1}(\mathfrak{C}_k)$ implies $h_\epsilon(x) = m(x)$.

10.6   The following two lemmas are important for certain cohomological results.

LEMMA 1:

Given:

$1°$ *Closed orientable 2-manifolds,* X *and* $\mathfrak{X}$.

$2°$ *Mappings,* $f_1:X \rightrightarrows \mathfrak{X}$ *and* $f_2:X \rightrightarrows \mathfrak{X}$.

$3°$ $\mathfrak{G}$ *is the interior of a 2-cell on* $\mathfrak{X}$ *such that:*

(a) $f_1^{-1}(\mathfrak{G}) = f_2^{-1}(\mathfrak{G})$.

(b) If $x \in f_1^{-1}(\mathfrak{G})$, then $f_1(x) = f_2(x)$.

Conclusion: $f_1^*:H^2(\mathfrak{X}) \to H^2(X)$ *and* $f_2^*:H^2(\mathfrak{X}) \to H^2(X)$ *are identical.*

Proof: If $G = f_1^{-1}(\mathfrak{G}) = f_2^{-1}(\mathfrak{G})$ and $\underline{f}_i = f_i|G$, $i = 1, 2$,   then

$\underline{f}_1 = \underline{f}_2 \equiv g$. Now consider the following diagram for $i = 1, 2$:

$$
\begin{array}{ccc}
H^2(G) & \xrightarrow{\;\;\;\;j\;\;\;\;} & H^2(X) \\[4pt]
{\scriptstyle g^*}\big\uparrow & & \big\uparrow{\scriptstyle f_i^*} \\[4pt]
H^2(\mathfrak{G}) & \xrightarrow{\;\;\;\;\mathfrak{k}\;\;\;\;} & H^2(\mathfrak{X})
\end{array}
$$

Commutativity holds, and one has

$$f_i^* \mathfrak{k} = j g^*, \; i = 1, 2.$$

But $\mathfrak{k}$ is an isomorphism onto by 9.2. Therefore:

$$f_i^* = j g^* \mathfrak{k}^{-1}, \; i = 1, 2,$$

and the lemma is proved.

LEMMA 2:

Given:

    $1^{\circ}.$ *Closed orientable 2-manifolds, X and $\mathfrak{X}$.*

    $2^{\circ}.$ *A mapping, $f:X \rightrightarrows \mathfrak{X}$.*

    $3^{\circ}.$ $\mathfrak{G}$ *is the interior of a 2-cell on $\mathfrak{X}$ such that if* $x \in f^{-1}(\mathfrak{G})$, *then* $x = f^{-1}f(x)$.

Conclusion: $f^*:H^2(\mathfrak{X}) \approx H^2(X)$.

Proof: Let $G = f^{-1}(\mathfrak{G})$ and use the diagram of lemma 1, replacing f_i by f. Now both j and $\mathfrak{k}$ are isomorphisms onto. But g is a homeomorphism, hence g^* is an isomorphism onto, and the proof is complete.

10.7   THEOREM:

Given:

    $1^{\circ}.$ *Closed orientable 2-manifolds, X and $\mathfrak{X}$.*

    $2^{\circ}.$ *A monotone mapping, $m:X \rightrightarrows \mathfrak{X}$.*

Conclusion: $m^*:H^2(\mathfrak{X}) \approx H^2(X)$.

Proof: By 10.4 there is a 2-cell on $\mathfrak{X}$ with interior $\mathfrak{G}$ and a mapping $\mu:X \rightrightarrows \mathfrak{X}$ such that: $\mu(x) = m(x)$ for $x \in X - m^{-1}(\mathfrak{G})$, and $x = \mu^{-1}\mu(x)$ if $x \in m^{-1}(\mathfrak{G})$. Hence μ satisfies the conditions of lemma 1 relative to the interior of any 2-cell in $\mathfrak{X} - \mathfrak{G}$ and the conditions of lemma 2 relative to $\mathfrak{G}$. Hence $m^* = \mu^*$ for the 2-dimensional cohomology groups, and $\mu^*:H^2(\mathfrak{X}) \approx H^2(X)$.

10.8   An immediate consequence of 10.7 is extremely useful in the applications.

THEOREM: *If $m:G \rightrightarrows \mathfrak{G}$ is a compact monotone mapping, where G and $\mathfrak{G}$ are open 2-cells, then $m^*:H^2(\mathfrak{G}) \approx H^2(G)$.*

Proof: If $\check{G}$ $[\check{\mathfrak{G}}]$ is a compactification of G $[\mathfrak{G}]$ by the addition of a point $\check{g}$ $[\check{\mathfrak{g}}]$, then $\check{m}:\check{G} \rightrightarrows \check{\mathfrak{G}}$ is a monotone mapping. Moreover, $\check{G}$ and $\check{\mathfrak{G}}$ are 2-spheres. Hence $\check{m}^*:H^2(\check{\mathfrak{G}}, \check{\mathfrak{g}}) \approx H^2(\check{G}, \check{g})$ by theorem 10.7. But this is simply the conclusion of the theorem:

10.9   Some results concerning the possibility of extending certain homomorphisms are of utmost importance in this paper and are now considered.

EXTENSION THEOREM (Youngs [16]):

Given:

1º. *X and $\mathfrak{X}$ are homeomorphic 2-manifolds with boundary.*

2º. *$\dot{h}:\dot{X} \approx \dot{\mathfrak{X}}$.*

Conclusion:

3º. *If X is non-orientable, then $\dot{h}$ can be extended to a homeomorphism $h:X \approx \mathfrak{X}$.*

4º. *If X is orientable, then $\dot{h}$ can be extended to a homeomorphism $h:X \approx \mathfrak{X}$ if and only if $\dot{h}$ carries a concordant orientation of $\dot{X}$ into a concordant orientation of $\dot{\mathfrak{X}}$* (see 9.8).

CHAPTER III

COMPARABILITY AND CONSISTENCY

The first chapter has indicated that the solution of the representation problem will employ the concept of *normal regions* (see 2.12). It is the object of this chapter to consider the analytic and algebraic properties of normal regions which lead to the concepts of comparability and consistency, concepts which are basic in the solution of the problem.

11. Hyper elements

The ordinary cyclic elements of a mantoid do not yield a fine enough decomposition of the mantoid for the purposes of the representation problem, and for this reason hyper elements are introduced. To obtain a rough picture of the situation, consider a space obtained from two tangent 2–spheres by identifying two points, each 2–sphere containing exactly one of the points. This space is a mantoid, since it can be realized as the monotone image of a torus. However, the space has exactly one true cyclic element, namely itself. On the other hand, it is the union of two 2–spheres, and for the purposes of the representation problem it is important to deal with these 2–spheres, neither a cyclic element but each a hyper element, as will be seen in a moment.

11.1   THEOREM: *If $\mathfrak{X}$ is a mantoid while $(\theta_1, \mathfrak{Y}_1, \mathfrak{X})$ and $(\theta_2, \mathfrak{Y}_2, \mathfrak{X})$ are suitable systems, then for any true cyclic element, $\mathfrak{C}_1$, of $\mathfrak{Y}_1$ there is a unique true cyclic element, $\mathfrak{C}_2$, of $\mathfrak{Y}_2$ such that $\theta_1(\mathfrak{C}_1) = \theta_2(\mathfrak{C}_2)$.*

Proof: Let $\mathfrak{x} \in \mathfrak{P}$ if either $\theta_1^{-1}(\mathfrak{x})$ or $\theta_2^{-1}(\mathfrak{x})$ is non–degenerate, and define $\Omega_i = \theta_i^{-1}(\mathfrak{P})$, $i = 1, 2$. It is easy to see that $\mathfrak{P}$ and Ω_i are finite, $i = 1, 2$, and that $\theta_i^{-1}\theta_i(\mathfrak{y}_i) = \mathfrak{y}_i$ if $\mathfrak{y}_i \in \mathfrak{Y}_i - \Omega_i$. (Compare with 6.1.)

Since $\mathfrak{Y}_1$ is a generalized cactoid, $\mathfrak{C}_1$ is a closed 2–manifold. Consequently, $(\mathfrak{C}_1 - \Omega_1)$ is cyclicly connected; that is to say, $[(\mathfrak{C}_1 - \Omega_1) - \mathfrak{y}_1]$ is connected for every $\mathfrak{y}_1 \in (\mathfrak{C}_1 - \Omega_1)$. Hence $\theta_2^{-1}\theta_1(\mathfrak{C}_1 - \Omega_1)$ is cyclicly connected and therefore lies in a unique true cyclic element, $\mathfrak{C}_2$, of $\mathfrak{Y}_2$. It will be shown that $\theta_2(\mathfrak{C}_2) = \theta_1(\mathfrak{C}_1)$.

The set $(\mathfrak{C}_2 - \Omega_2)$ is cyclicly connected, since $\mathfrak{C}_2$ is a 2-manifold. Hence $\theta_1^{-1}\theta_2(\mathfrak{C}_2 - \Omega_2)$ is cyclicly connected. But $\theta_1^{-1}\theta_2(\mathfrak{C}_2 - \Omega_2)$ has a non-degenerate intersection with $\mathfrak{C}_1$ and hence $\mathfrak{C}_1$ is the unique true cyclic element of $\mathfrak{Y}_1$ containing $\theta_1^{-1}\theta_2(\mathfrak{C}_2 - \Omega_2)$. It is now a simple matter to see that $\theta_1(\mathfrak{C}_1 - \Omega_1) = \theta_2(\mathfrak{C}_2 - \Omega_2)$ and consequently, $\theta_1(\mathfrak{C}_1) = \theta_1(\overline{\mathfrak{C}_1 - \Omega_1})$ $= \overline{\theta_1(\mathfrak{C}_1 - \Omega_1)} = \overline{\theta_2(\mathfrak{C}_2 - \Omega_2)} = \theta_2(\overline{\mathfrak{C}_2 - \Omega_2}) = \theta_2(\mathfrak{C}_2)$.

11.2   *Definition:* A set, $\mathfrak{L}$, in a mantoid, $\mathfrak{X}$, is called a *hyper element* of $\mathfrak{X}$ if and only if there is a suitable system, $(\theta, \mathfrak{Y}, \mathfrak{X})$, and a true cyclic element, $\mathfrak{C}$, of $\mathfrak{Y}$ such that $\theta(\mathfrak{C}) = \mathfrak{L}$.

Remark: The theorem has shown that the hyper elements of $\mathfrak{X}$ are independent of the suitable system, $(\theta, \mathfrak{Y}, \mathfrak{X})$.

It is observed that a hyper element of a mantoid, $\mathfrak{X}$, is either a 2-manifold or the image of a 2-manifold under a mapping which can be factored into a finite number of 2-point identifications.

11.3   The hyper elements of a mantoid $\mathfrak{X}$ could also have been defined in an intrinsic manner by using the theory of k-cyclic elements (see Youngs [13]). The definition of the k-cyclic elements of a Peano space is analagous to a definition which may be employed in dealing with true cyclic elements and provides a finer decomposition of the space. For a mantoid, $\mathfrak{X}$, it may be shown that there is an integer, k_0, such that if k_1 and k_2 are integers not less than k_0, then the k_1- and k_2-cyclic elements coincide. The hyper elements of $\mathfrak{X}$ are precisely the k_0-cyclic elements of $\mathfrak{X}$. This approach, however, is not adopted in this paper due to the fact that the earlier definition in terms of suitable systems is more in line with the results of Steenrod and Roberts.

12.   Normal and simple regions

Normal regions, defined in 2.12, are important principally because of theorem 16.3 which shows the connection between normal and canonical regions.

12.1   THEOREM:

Given:

1°. *A normal region,* $\mathfrak{U}$, *in a mantoid,* $\mathfrak{X}$.

2°. $\mathfrak{R}$ *a component of* $\dot{\mathfrak{U}}$.

3°. $\epsilon > 0$.

Conclusion: *There is a region,* $\mathfrak{R}$, *of* $\mathfrak{X}$ *in* $\mathfrak{U}$ *such that:*

4°. $\overline{\mathfrak{R}} \cap \dot{\mathfrak{U}} = \mathfrak{R}$.

5°. $\dot{\mathfrak{R}}$ *has two components, and if* $\mathfrak{R}$ *is a point [Jordan curve],* *each component of* $\dot{\mathfrak{R}}$ *is a point [Jordan curve].*

6°. $\overline{\mathfrak{U} - \mathfrak{R}} \cap \mathfrak{R} = 0$.

7°. *No point of* $\mathfrak{R}$ *is at a distance greater than* ϵ *from some point of* $\mathfrak{R}$.

Proof: *Case I.* $\mathfrak{R}$ is an end point of $\overline{\mathfrak{U}}$. There is a cut point, $\mathfrak{p}$, of $\overline{\mathfrak{U}}$ with the following properties:

(i) The component, $\mathfrak{G}$, of $\overline{\mathfrak{U}} - \mathfrak{p}$ containing $\mathfrak{R}$ contains no point of $\dot{\mathfrak{U}} - \mathfrak{R}$.

(ii) $d(\mathfrak{G}) < \epsilon$.

The required set, $\mathfrak{R}$, is $(\mathfrak{G} - \mathfrak{R})$.

Case II. $\mathfrak{R}$ is a Jordan curve. Then there is a 2–cell, $\mathfrak{C}$, which is a true cyclic element of $\overline{\mathfrak{U}}$, and $\mathfrak{R}$ is the boundary of $\mathfrak{C}$.

Select a sequence, $\{\mathfrak{I}_n\}$, of Jordan curves, $\mathfrak{I}_n$, on $\mathfrak{C}$, $n = 1, 2, 3, \cdots$, with the following properties:

(i) $\mathfrak{I}_n \cap \mathfrak{R} = 0$, $n = 1, 2, 3, \cdots$.

(ii) $Lim\ \mathfrak{I}_n = \mathfrak{R}$.

(iii) If $\mathfrak{G}_n^*$ is the component of $\mathfrak{C} - \mathfrak{I}_n$ containing $\mathfrak{R}$, then $\mathfrak{G}_n^* \supset \mathfrak{G}_{n+1}^*$, $n = 1, 2, 3, \cdots$.

Each component, $\mathfrak{G}$, of $\overline{\mathfrak{U}} - \mathfrak{C}$ has a degenerate frontier relative to the Peano space $\overline{\mathfrak{U}}$. The frontier must, of course, lie in $\mathfrak{C}$; it is to be shown that it does not lie in $\mathfrak{R}$. If the frontier of $\mathfrak{G}$ is in $\mathfrak{R}$, then $\overline{\mathfrak{U}} - \mathfrak{R}$ is not connected. Since $\overline{\mathfrak{U}} - \mathfrak{R} \supset \mathfrak{U}$ and $\mathfrak{U}$ is a region, there is a component, $\mathfrak{G}^*$, of $\overline{\mathfrak{U}} - \mathfrak{R}$ containing $\mathfrak{U}$. Since $\mathfrak{R}$ is a component of $\dot{\mathfrak{U}}$, it follows that $\mathfrak{G}^* = \overline{\mathfrak{U}} - \mathfrak{R}$. Hence $\overline{\mathfrak{U}} - \mathfrak{R}$ is connected and each component, $\mathfrak{G}$, of $\overline{\mathfrak{U}} - \mathfrak{C}$ is such that its frontier relative to $\overline{\mathfrak{U}}$ is in $\mathfrak{C} - \mathfrak{R}$.

Let $\mathfrak{G}_n$ be the union of $\mathfrak{G}_n^*$ with those components of $\overline{\mathfrak{U}} - \mathfrak{C}$ having frontiers, relative to $\overline{\mathfrak{U}}$, in $\mathfrak{G}_n^*$, $n = 1, 2, 3, \cdots$. Since each of these components is open in $\bar{\mathfrak{U}}$, the set $\mathfrak{S}_n$ is open. Suppose d_n is the maximum of the diameters of the components of $\bar{\mathfrak{U}} - \mathfrak{C}$ in $\mathfrak{S}_n$, $n = 1, 2, 3,$ $\cdots$. It will be shown that $d_n \to 0$. If this is not the case, then since no component of $\bar{\mathfrak{U}} - \mathfrak{C}$ has its frontier in $\mathfrak{R}$, an infinite number of the components of $\bar{\mathfrak{U}} - \mathfrak{C}$ have diameters larger than some positive number. But this is impossible, since $\bar{\mathfrak{U}}$ is a Peano space.

If $\eta = min \; \rho\{\mathfrak{x}, \mathfrak{y}\}$, where $\mathfrak{x} \in \mathfrak{R}$ and $\mathfrak{y} \in \dot{\mathfrak{U}} - \mathfrak{R}$, then $\eta > 0$ since $\mathfrak{R}$ is a component of $\dot{\mathfrak{U}}$. Now there is clearly an integer n such that $\mathfrak{S}_n \cap \dot{\mathfrak{U}} = \mathfrak{R}$ and each point of $\mathfrak{S}_n$ is within ϵ of some point of $\mathfrak{R}$. The required set $\mathfrak{R}$ is simply $(\mathfrak{S}_n - \mathfrak{R})$.

12.2   THEOREM:

      Given:

           1°   *A mantoid,* $\mathfrak{X}$.

           2°   *A normal region,* $\mathfrak{U}$, *in* $\mathfrak{X}$.

           3°   *A true cyclic element,* $\mathfrak{C}$, *of* $\bar{\mathfrak{U}}$.

           4° $\mathfrak{C} \cap \dot{\mathfrak{U}} \neq 0$.

      Conclusion: $\mathfrak{C}$ *is a 2-cell bounded by* $\mathfrak{C} \cap \dot{\mathfrak{U}}$.

Proof: No point component of $\dot{\mathfrak{U}}$ can be on $\mathfrak{C}$, since each of these is an end point of $\dot{\mathfrak{U}}$. Suppose $\mathfrak{C}$ has a point on a Jordan curve, $\mathfrak{J}$, in $\dot{\mathfrak{U}}$. The true cyclic element, $\mathfrak{C}$, of $\bar{\mathfrak{U}}$ containing $\mathfrak{J}$ is a 2-cell bounded by $\mathfrak{J}$ (see 2.12). It will be shown that $\mathfrak{C} = \mathfrak{C}$.

This is certainly the case if $\mathfrak{C} \cap \mathfrak{C}$ is non-degenerate. Suppose, therefore, that $\mathfrak{C} \cap \mathfrak{C}$ is a single point, $\mathfrak{x}$. Now $\mathfrak{C} \supset \mathfrak{J}$, $\mathfrak{C} \cap J \neq 0$, and $\mathfrak{C} \cap \mathfrak{C} = \mathfrak{x}$, hence $\mathfrak{x} \in \mathfrak{J}$. The set $(\mathfrak{C} - \mathfrak{x})$ is connected and lies in $\bar{\mathfrak{U}} - \mathfrak{C}$, consequently there is a component of $\bar{\mathfrak{U}} - \mathfrak{C}$ with frontier $\mathfrak{x} \in \mathfrak{J}$. But this is impossible. Hence $\mathfrak{C} = \mathfrak{C}$ and is bounded by $\mathfrak{J}$.

The fact that $\mathfrak{J} = \mathfrak{C} \cap \dot{\mathfrak{U}}$ is easy to show since no point component of $\dot{\mathfrak{U}}$ can lie on $\mathfrak{C}$, and if a Jordan curve component, $\mathfrak{J}'$, of $\dot{\mathfrak{U}}$ (other than $\mathfrak{J}$) is such that $\mathfrak{J}' \cap \mathfrak{C} \neq 0$, then it has been shown above that $\mathfrak{C}$ is bounded by $\mathfrak{J}'$. Consequently, the 2-cell, $\mathfrak{C}$, is bounded by two Jordan curves whose intersection is vacuous. This is impossible.

12.3   THEOREM:

      Given:

1° *A mantoid,* $\mathfrak{X}$, *and a monotone mapping,* $m: X \rightrightarrows \mathfrak{X}$, *where* X *is a closed 2-manifold.*

2° *A normal region,* $\mathfrak{U}$, *in* $\mathfrak{X}$.

3° $U = m^{-1}(\mathfrak{U})$.

Conclusion: *U is a canonical region, and if K_1, $\cdots$, K_r are the components of $\dot{U}$, then $m(K_1)$, $\cdots$, $m(K_r)$ are the components of $\dot{\mathfrak{U}}$.*

The theorem follows directly on combining 12.1 with 6.9.

COROLLARY: *With the hypothesis of the theorem, if $\bar{m} = m\,|\,\bar{U}$, then $\bar{m}: (\bar{U},\ \dot{U}) \rightrightarrows (\bar{\mathfrak{U}},\ \dot{\mathfrak{U}})$.*

__12.4__   THEOREM: *If $\mathfrak{U}$ is a normal region in a mantoid, $\mathfrak{X}$, then there is a sequence, $\{\mathfrak{B}_n\}$, of normal regions, $\mathfrak{B}_n$, $n = 1, 2, 3, \cdots$, such that:*

1° $\mathfrak{B}_1 \subset \bar{\mathfrak{B}}_1 \subset \mathfrak{B}_2 \subset \bar{\mathfrak{B}}_2 \subset \cdots \subset \mathfrak{U}$.

2° $\bigcup \mathfrak{B}_n = \mathfrak{U}$.

3° *$\dot{\mathfrak{B}}_n$ and $\dot{\mathfrak{U}}$ have the same number of components.*

The proof is an exercise in the methods employed in 12.1.

COROLLARY:

Given:

1° *Two monotone mappings, $m_1: X_1 \rightrightarrows \mathfrak{X}$ and $m_2: X_2 \rightrightarrows \mathfrak{X}$, where X_1 and X_2 are closed 2-manifolds.*

2° *A normal region, $\mathfrak{U}$, in $\mathfrak{X}$.*

3° *A 2-manifold, M_i, associated with the canonical region $U_i = m_i^{-1}(\mathfrak{U})$, $i = 1, 2$.*

Conclusion: *There is a sequence, $\{\mathfrak{B}_n\}$, of normal regions, $\mathfrak{B}_n$, $n = 1, 2, 3, \cdots$, such that conditions 1°, 2°, and 3° of the theorem are satisfied and, in addition, $m_i^{-1}(\mathfrak{B}_n) \supset M_i$, $i = 1, 2;\ n = 1, 2, 3, \cdots$.*

__12.5__   *Definition:* A set, $\mathfrak{U}$, in a mantoid, $\mathfrak{X}$, is said to be a *simple region* if and only if:

1° $\mathfrak{U}$ is a normal region (see 2.12).

2° If $\mathfrak{C}$ is a true cyclic element of $\bar{\mathfrak{U}}$, then $\mathfrak{C}$ is either a hyper element of $\mathfrak{X}$ or a 2-cell.

It will be noticed that each true cyclic element of $\bar{\mathfrak{U}}$ is a 2-cell, a closed 2-manifold, or a set which is the image of a 2-manifold under a mapping which can be factored into a finite number of 2–point identifications. More-over, *a true cyclic element of* $\bar{\mathfrak{U}}$ *is a 2–cell if and only if it is bounded by a Jordan curve component of* $\dot{\mathfrak{U}}$.

For the purposes of this paper, the following special definitions will be useful.

Definition: A space, $\mathfrak{X}$, is said to be a *finite mantoid* if and only if:

$1\overset{\circ}{.}$ $\mathfrak{X}$ is a mantoid.

$2\overset{\circ}{.}$ $\mathfrak{X}$ has only a finite number of hyper elements, $\mathfrak{L}_1, \cdots, \mathfrak{L}_l$.

$3\overset{\circ}{.}$ There are a finite number of points, $\mathfrak{p}_1, \cdots, \mathfrak{p}_p$, such that $[\mathfrak{X} - (\bigcup \mathfrak{L}_k \cup \bigcup \mathfrak{p}_k)] = 0$ or has a finite number of components, each an open arc.

Definition: A space, $\mathfrak{X}$, is said to be a *cluster* if and only if it is a mantoid with a finite number of hyper elements, $\mathfrak{L}_1, \cdots, \mathfrak{L}_l$, and $\mathfrak{X} = \bigcup \mathfrak{L}_k$.

A cluster is therefore a special type of finite mantoid; one, in fact, with $p = 0$ and $(\mathfrak{X} - \bigcup \mathfrak{L}_k) = 0$.

<u>12.6</u>   *Definition:* If $\mathfrak{X}$ is a mantoid and $\mathfrak{x} \in \mathfrak{X}$, then a set, $\mathfrak{U}$, is said to be a *simple neighborhood of* $\mathfrak{x}$ if and only if $\mathfrak{U}$ is a simple region in $\mathfrak{X}$ containing $\mathfrak{x}$, such that $\mathfrak{U} - \mathfrak{x}$ has a finite number of components, $\mathfrak{S}$, and $\bar{\mathfrak{S}}$ is either an arc or a 2–cell.

It is clear that if $\mathfrak{U}$ is a simple neighborhood of a point, $\mathfrak{x}$, in a cluster, $\mathfrak{X}$, while $\mathfrak{S}$ is a component of $\mathfrak{U} - \mathfrak{x}$, then $\bar{\mathfrak{S}}$ is a 2–cell. On the other hand, it is easy to exhibit an example of a mantoid $\mathfrak{X}$ (in fact a cactoid) such that if $\mathfrak{x} \in \mathfrak{X}$, then no simple neighborhood of $\mathfrak{x}$ exists. The next result, an existence theorem, serves to clarify this situation. The proof is an exercise involving the use of theorem 6.1.

THEOREM: *If* $\mathfrak{X}$ *is a finite mantoid,* $\mathfrak{x} \in \mathfrak{X}$, *and* $\epsilon > 0$, *then there is a simple neighborhood of* $\mathfrak{x}$ *with diameter less than* ϵ.

13.   Cohomology on hyper elements

This section prepares the ground for the definition of consistency to be offered in section 18. The principal theorem is that of 13.7 and should be compared with the situation which obtains if one is concerned with a monotone mapping $m:X \rightrightarrows \mathfrak{X}$ where X is a 2-sphere. In this case $\mathfrak{X}$ is a cactoid and, in so far as the representation problem is concerned, it is possible to deal with the largest simple region in $\mathfrak{X}$, namely $\mathfrak{X}$ itself. Now if $\mathfrak{L}$ is any true cyclic element of $\mathfrak{X}$ while $r:\mathfrak{X} \rightrightarrows \mathfrak{L}$ is the monotone retraction and $\rho = rm$, then $\rho^*:H^2(\mathfrak{L}) \approx H^2(\mathfrak{X})$. This fact is proved, using the concept of the degree of a mapping, in Youngs [16]. Theorem 13.7 therefore illustrates some of the additional complications which arise if X is a general closed 2-manifold rather than a 2-sphere.

13.1 THEOREM:

Given:

1°. *A monotone mapping* $m:X \rightrightarrows \mathfrak{X}$ *where* X *is a closed 2-manifold.*

2°. *A hyper element,* $\mathfrak{L}$, *of* $\mathfrak{X}$.

3°. *The true cyclic element,* $\mathfrak{T}$, *of* $\mathfrak{X}$ *containing* $\mathfrak{L}$.

4°. *The monotone retraction,* $r:\mathfrak{X} \rightrightarrows \mathfrak{T}$.

5°. *The set,* $\mathfrak{F}(\mathfrak{L}, m)$, *of points,* x, *in* $\mathfrak{L}$ *such that* $R(m^{-1}r^{-1}(x)) \neq 0$. (See 6.3)

Conclusion:

6°. $\mathfrak{F}(\mathfrak{L}, m)$ *is a finite set.*

7°. $\mathfrak{L} - \mathfrak{F}(\mathfrak{L}, m)$ *is homeomorphic to a closed 2-manifold with a finite set removed.*

8°. *If* $\mathfrak{U}$ *is contained in* $\mathfrak{L} - \mathfrak{F}(\mathfrak{L}, m)$ *and is the interior of a 2-cell, then* $r^{-1}(\mathfrak{U})$ *is a normal region in* $\mathfrak{X}$ *and* $m^{-1}r^{-1}(\mathfrak{U})$ *is an open 2-cell in* X.

9°. *If* $\mathfrak{U}$ *is the interior of a 2-cell in* $\mathfrak{L}$ *and* $m^{-1}r^{-1}(\mathfrak{U})$ *is an open 2-cell in* X, *then* $\mathfrak{U} \cap \mathfrak{F}(\mathfrak{L}, m) = 0$.

Proof: If $\bar{m} = rm$, then $\bar{m}$ is uniquely determined by $\mathfrak{L}$ since $\mathfrak{T}$ is uniquely determined by $\mathfrak{L}$ and r is uniquely determined by $\mathfrak{T}$. Now $\bar{m}:X \rightrightarrows \mathfrak{T}$ is a monotone mapping from a closed 2-manifold, hence by 6.8 there is:

(i) A closed 2-manifold, Y (which is orientable if X is orientable).

(ii) A suitable system, $(\theta, \mathfrak{Y}, \mathfrak{T})$, with exceptional sets $\mathfrak{P}$ in $\mathfrak{T}$ and Ω in $\mathfrak{Y}$.

(iii) A monotone mapping, $\mu : Y \rightrightarrows \mathfrak{Y}$, and a homeomorphism, $h : [X - \bar{m}^{-1}(\mathfrak{P})] \approx [Y - \mu^{-1}(\Omega)]$, such that $\bar{m}(x) = \theta \mu h(x)$ for $x \in [X - \bar{m}^{-1}(\mathfrak{P})]$.

The last statement implies that if $x \in \mathfrak{X} - \mathfrak{P}$, then $\bar{m}^{-1}(x) = h^{-1}\mu^{-1}\theta^{-1}(x)$.

Since $(\theta, \mathfrak{Y}, \mathfrak{X})$ is a suitable system, the space $\mathfrak{Y}$ is a generalized cactoid and, by 11.1, there is a unique true cyclic element, $\mathfrak{C}$, of $\mathfrak{Y}$ such that $\theta(\mathfrak{C}) = \mathfrak{L}$. Consider the monotone retraction $\rho : \mathfrak{Y} \rightrightarrows \mathfrak{C}$ and let $\bar{\mu} = \rho\mu$. Then $\bar{\mu} : Y \rightrightarrows \mathfrak{C}$ is a monotone mapping from a closed 2-manifold into another. Hence by 6.6 there is a finite set $\mathfrak{A}$ such that if $\mathfrak{y} \in \mathfrak{C} - \mathfrak{A}$, then $R(\bar{\mu}^{-1}(\mathfrak{y})) = 0$ and if $\mathfrak{y} \in \mathfrak{A}$, then $R(\bar{\mu}^{-1}(\mathfrak{y})) \neq 0$.

If $\mathfrak{B} = \rho(\Omega)$, then $\mathfrak{B}$ is a finite set since Ω is finite. If $\mathfrak{y} \in \mathfrak{C} - \mathfrak{B}$, then $\rho^{-1}(\mathfrak{y}) = \mathfrak{y}$; for otherwise there is a component, $\mathfrak{G}$, of $\mathfrak{Y} - \mathfrak{C}$ such that $\dot{\mathfrak{G}} = \rho(\mathfrak{G}) = \mathfrak{y}$, hence $\mathfrak{G} \cap \Omega = 0$ and it follows that $\theta(\mathfrak{y})$ cuts the cyclicly connected space $\mathfrak{X}$.

Therefore, if $\mathfrak{y} \in [\mathfrak{C} - (\mathfrak{A} \cup \mathfrak{B})]$, then $\bar{\mu}^{-1}(\mathfrak{y}) = \mu^{-1}\rho^{-1}(\mathfrak{y}) = \mu^{-1}(\mathfrak{y})$ and hence $R(\mu^{-1}(\mathfrak{y})) = 0$. Moreover, $\mathfrak{y} \in (\mathfrak{C} - \Omega)$, and hence $\mathfrak{y} = \theta^{-1}\theta(\mathfrak{y})$. Consequently, if $x \in [\mathfrak{L} - \theta(\mathfrak{A} \cup \mathfrak{B})]$, then $\theta^{-1}(x)$ is a single point in $[\mathfrak{C} - (\mathfrak{A} \cup \mathfrak{B})]$ and $R(\mu^{-1}\theta^{-1}(x)) = 0$. On the other hand, $[\mathfrak{L} - \theta(\mathfrak{A} \cup \mathfrak{B})] \subset (\mathfrak{X} - \mathfrak{P})$ hence $\bar{m}^{-1}(x) = h^{-1}\mu^{-1}\theta^{-1}(x)$. Now consider any open set, G^*, containing $\bar{m}^{-1}(x)$. If $G = G^* - \bar{m}^{-1}(\mathfrak{P})$, then G is also an open set containing $\bar{m}^{-1}(x)$. Hence $h(G)$ is an open set in Y containing $\mu^{-1}\theta^{-1}(x)$. Since $R(\mu^{-1}\theta^{-1}(x)) = 0$ there is a 2-cell, C, in $h(G)$ such that $\mu^{-1}\theta^{-1}(x) \subset C^\circ$ by 6.4. Therefore, $h^{-1}(C)$ is a 2-cell in G^* containing $\bar{m}^{-1}(x)$ in its interior, and $R(\bar{m}^{-1}(x)) = 0$ by 6.4.

To summarize, it has been shown that $x \in [\mathfrak{L} - \theta(\mathfrak{A} \cup \mathfrak{B})]$ implies $R(\bar{m}^{-1}(x)) = 0$. It will be shown that $x \in \theta(\mathfrak{A} \cup \mathfrak{B})$ implies $R(\bar{m}^{-1}(x)) \neq 0$.

First suppose that $x \in \theta(\mathfrak{B})$. Select any $\mathfrak{y} \in \mathfrak{B} \cap \theta^{-1}(x)$. If $\mathfrak{y} \in \Omega$, then x is a local cut point of $\mathfrak{X}$ since there is a point $\mathfrak{y}_1 \in \Omega$ such that $\mathfrak{y}_1 \neq \mathfrak{y}$ and $\theta(\mathfrak{y}_1) = x$. If $\mathfrak{y} \notin \Omega$, then (since $\mathfrak{y} \in \rho(\Omega)$) there is a component, $\mathfrak{G}$, of $\mathfrak{Y} - \mathfrak{C}$ such that $\mathfrak{G} \cap \Omega \neq 0$ and $\dot{\mathfrak{G}} = \mathfrak{y}$; and now it follows again that x is a local cut point of $\mathfrak{X}$. Consequently, if $x \in \theta(\mathfrak{B})$ there is a connected open set, $\mathfrak{G}$, containing x and having the property that $\mathfrak{G} - x$ is not connected. In view of monotoneity, $\bar{m}^{-1}(\mathfrak{G})$ is a connected open set containing $\bar{m}^{-1}(x)$, and $[\bar{m}^{-1}(\mathfrak{G}) - \bar{m}^{-1}(x)]$ is not connected. If

$R(\bar{m}^{-1}(x)) = 0$, then there is a 2-cell, C, such that $\bar{m}^{-1}(x) \subset C^{\circ} \subset C \subset m^{-1}(\mathfrak{G})$; consequently, $[\bar{m}^{-1}(\mathfrak{G}) - \bar{m}^{-1}(x)]$ is connected. (See 6.4.) This contradiction shows that $x \in \theta(\mathfrak{B})$ implies $R(\bar{m}^{-1}(x)) \neq 0$.

If $x \in \theta(\mathfrak{A}) - \theta(\mathfrak{B})$, then $\theta^{-1}(x)$ is a single point, $\mathfrak{h}$, in $\mathfrak{A} - \mathfrak{B}$. If $R(\bar{m}^{-1}(x)) = 0$, then it follows that $R(\bar{\mu}^{-1}(\mathfrak{h})) = 0$ which is impossible since $\mathfrak{h} \in \mathfrak{A}$.

Consequently, the set $\mathfrak{F}(\mathfrak{L}, m)$ of 5° is precisely $\theta(\mathfrak{A} \cup \mathfrak{B})$ and is finite since $\mathfrak{A}$ and $\mathfrak{B}$ are finite. This proves item 6°.

The mapping $\theta \mid (\mathfrak{Y} - \mathfrak{Q})$ is a homeomorphism from $(\mathfrak{Y} - \mathfrak{Q})$ onto $(\mathfrak{T} - \mathfrak{P})$, and $\theta^{-1}(\mathfrak{T} - \mathfrak{P}) = (\mathfrak{Y} - \mathfrak{Q})$. Hence $\theta \mid [\mathfrak{E} - (\mathfrak{A} \cup \mathfrak{B})]$ is a homeomorphism from $[\mathfrak{E} - (\mathfrak{A} \cup \mathfrak{B})]$ onto $[\mathfrak{L} - \mathfrak{F}(\mathfrak{L}, m)]$, and $\theta^{-1}[\mathfrak{L} - \mathfrak{F}(\mathfrak{L}, m)] = [\mathfrak{E} - (\mathfrak{A} \cup \mathfrak{B})]$. But $\mathfrak{E}$ is a 2-manifold and $(\mathfrak{A} \cup \mathfrak{B})$ is a finite set; consequently item 7° has been proved.

If $\mathfrak{U}$ lies in $[\mathfrak{L} - \mathfrak{F}(\mathfrak{L}, m)]$ and is the interior of a 2-cell, then $\mathfrak{U}$ is a simple region in $\mathfrak{T}$ with respect to $\mathfrak{T}$, such that $\dot{\mathfrak{U}}$ consists of a single Jordan curve. It is a simple matter to see that $r^{-1}(\mathfrak{U})$ is a normal region in $\mathfrak{X}$. If $U = \bar{m}^{-1}(\mathfrak{U})$, then $\dot{U}$ has a single component and $\bar{m}(\dot{U}) = \dot{\mathfrak{U}}$.

Let $\mathfrak{M}$ be the closed 2-manifold obtained from $\bar{U}$ by contracting $\dot{U}$ to a point, and suppose $\phi : \bar{U} \rightrightarrows \mathfrak{M}$ is the associated mapping. Let $\mathfrak{N}$ be the 2-sphere obtained from $\bar{\mathfrak{U}}$ by contracting $\dot{\mathfrak{U}}$ to a point, and suppose $\psi : \bar{\mathfrak{U}} \rightrightarrows \mathfrak{N}$ is the associated mapping. Define the monotone mapping $\gamma : \mathfrak{M} \rightrightarrows \mathfrak{N}$ by the formula $\gamma = \psi \bar{m} \phi^{-1}$.

If $\mathfrak{M}$ is not a 2-sphere, then by 6.5 there is at least one point, n, in $\mathfrak{N}$ such that $R(\gamma^{-1}(n)) \neq 0$. The point n clearly cannot be $\psi(\dot{\mathfrak{U}})$, hence $\psi^{-1}(n)$ is a point, x, in $\mathfrak{U}$ and so in $\mathfrak{L} - \mathfrak{F}(\mathfrak{L}, m)$. Hence $R(\bar{m}^{-1}(x)) = 0$. On the other hand, since $n \neq \psi(\dot{\mathfrak{U}})$ it follows that $R(\bar{m}^{-1}(x)) = 0$ if and only if $R(\gamma^{-1}(n)) = 0$. This contradiction shows that $\mathfrak{M}$ is a 2-sphere, hence $\mathfrak{M} - \phi(\dot{U})$ is an open 2-cell. Therefore U is also an open 2-cell; hence item 8° has been proved.

Suppose $\mathfrak{U}$ is the interior of a 2-cell in $\mathfrak{L}$ and $\bar{m}^{-1}(\mathfrak{U})$ is an open 2-cell in X. Since $\bar{m} : X \rightrightarrows \mathfrak{T}$ and $\bar{m}^{-1}(\mathfrak{U})$ is open, $\mathfrak{U}$ is open with respect to $\mathfrak{T}$. Let $x \in \mathfrak{U}$ and consider $\bar{m}^{-1}(x)$. If G is an open set containing $\bar{m}^{-1}(x)$ there is an open 2-cell, $\mathfrak{U}^*$, such that $x \in \mathfrak{U}^* \subset \bar{\mathfrak{U}}^* \subset \mathfrak{U}$, and $\bar{m}^{-1}(\mathfrak{U}^*) \subset G$. It is easy to see that $\bar{m}^{-1}(\mathfrak{U}^*)$ is an open 2-cell, whence it follows that there is a 2-cell, C, such that $\bar{m}^{-1}(x) \subset C^\circ \subset C \subset G$. Consequently, $R(\bar{m}^{-1}(x)) = 0$ by 6.4. Therefore $\mathfrak{U} \cap \mathfrak{F}(\mathfrak{L}, m) = 0$, and this completes the proof of the theorem.

<u>13.2</u>   Suppose that $m: X \rightrightarrows X$ is a monotone mapping where X is a closed 2-manifold. Let $\mathfrak{U}$ be a *simple* region in X and $\mathfrak{L}$ be any true cyclic element of $\bar{\mathfrak{u}}$. There is a unique hyper element, $\mathfrak{L}^*$, of X containing $\mathfrak{L}$; for if $\mathfrak{L}$ is not a 2-cell, $\mathfrak{L}^* = \mathfrak{L}$; and if $\mathfrak{L}$ is a 2-cell, then no finite set cuts $\mathfrak{L}$, and the statement follows. Let $\mathfrak{X}$ be the unique true cyclic element of X containing $\mathfrak{L}^*$, while $\bar{r}: X \rightrightarrows \mathfrak{X}$ and $r: \bar{\mathfrak{u}} \rightrightarrows \mathfrak{L}$ are the monotone retractions.

      THEOREM: *With the conventions above, the set* $[\mathfrak{L} - \{\mathfrak{F}(\mathfrak{L}^*, m) \cup r(\dot{\mathfrak{u}})\}]$ *is homeomorphic to a closed 2-manifold with a finite set removed; if* $G \subset [\mathfrak{L} - \{\mathfrak{F}(\mathfrak{L}^*, m) \cup r(\dot{\mathfrak{u}})\}]$ *and* G *is the interior of a 2-cell, then* $r^{-1}(G)$ *is a simple region in* X *and* $m^{-1}r^{-1}(G)$ *is an open 2-cell in* X; *if* G *is the interior of a 2-cell in* $\mathfrak{L}$ *and* $m^{-1}r^{-1}(G)$ *is an open 2-cell in* X, *then* $G \cap [\mathfrak{F}(\mathfrak{L}^*, m) \cup r(\dot{\mathfrak{u}})] = 0$.

      Proof: The first part of the conclusion is evident in view of 13.1, 7°, and the definition of a simple region.

      To prove the second part, notice that if G is a component of $\bar{\mathfrak{u}} - \mathfrak{L}$ whose frontier with respect to $\bar{\mathfrak{u}}$ is in G, then $G \cap \dot{\mathfrak{u}} = 0$ since otherwise $G \cap r(\dot{\mathfrak{u}}) \neq 0$. If G has a single point in common with $\mathfrak{X}$, then $G \subset \mathfrak{X}$ since $\mathfrak{X}$ is a cyclic element, and it follows that $G \cap \dot{\mathfrak{u}} \neq 0$. This proves that $G \subset (X - \mathfrak{X})$ and as $G \cap \dot{\mathfrak{u}} = 0$, it is also a component of $X - \mathfrak{X}$.

      Conversely, if G is a component of $X - \mathfrak{X}$ whose frontier is in G, then it follows that $G \subset \mathfrak{U}$ and hence is a component of $\bar{\mathfrak{u}} - \mathfrak{L}$.

      These two paragraphs show that $r^{-1}(G) = \bar{r}^{-1}(G)$ and hence by 13.1, 8°, $r^{-1}(G)$ is a simple region in X. Moreover $m^{-1}r^{-1}(G)$ is an open 2-cell in X.

      Suppose G is the interior of a 2-cell in $\mathfrak{L}$ and $m^{-1}r^{-1}(G)$ is an open 2-cell in X. If $\mathfrak{L}$ is a 2-cell, then it is clear that G cannot contain any point of the bounding Jordan curve of $\mathfrak{L}$ since G is the interior of a 2-cell in $\mathfrak{L}$. And now, for any $\mathfrak{L}$, it is easy to show that

$\mathfrak{G} \cap r(\dot{\mathfrak{u}}) \neq 0$ violates the statement that $m^{-1}r^{-1}(\mathfrak{G})$ is an open 2-cell in X. Consequently, $r^{-1}(\mathfrak{G}) = \bar{r}^{-1}(\mathfrak{G})$ and now $\mathfrak{G} \cap \mathfrak{F}(\mathfrak{Q}^*, m) = 0$ follows from theorem 13.1, 9°.

13.3   THEOREM:

Given:

1°. *A mantoid, X.*

2°. *A hyper element, $\mathfrak{Q}$, of X.*

3°. *A suitable system, $(\theta, \mathfrak{Y}, X)$.*

4°. *The true cyclic element, $\mathfrak{S}$, of $\mathfrak{Y}$ such that $\theta(\mathfrak{S}) = \mathfrak{Q}$.*

Conclusion: *If $\bar{\theta} = \theta|\mathfrak{S}$, then $\bar{\theta}^*:H^2(\mathfrak{Q}) \approx H^2(\mathfrak{S})$.*

Proof: The theorem is obvious if $\bar{\theta}$ is a homeomorphism. To consider the remaining possibility, let $x \in \mathfrak{P}^*$ if and only if $x \in \mathfrak{Q}$ and $\bar{\theta}^{-1}(x)$ is non-degenerate. If $\mathfrak{Q}^* = \bar{\theta}^{-1}(\mathfrak{P}^*)$, then $\mathfrak{P}^*$ and $\mathfrak{Q}^*$ are finite sets. Consider the diagram below, where $\underline{\theta} = \bar{\theta}$.

$$
\begin{array}{ccc}
H^2(\mathfrak{S}) & \xleftarrow{\ \bar{\theta}^*\ } & H^2(\mathfrak{Q}) \\[1mm]
\big\uparrow a^* & & \big\uparrow b^* \\[1mm]
H^2(\mathfrak{S},\ \mathfrak{Q}^*) & \xleftarrow{\ \underline{\theta}^*\ } & H^2(\mathfrak{Q},\ \mathfrak{P}^*)
\end{array}
$$

Commutativity holds, while the homomorphisms a^* and b^* are isomorphisms onto by 7.8 and $\underline{\theta}^*$ by 7.6. Consequently, $\bar{\theta}^*$ is an isomorphism onto.

Remark: In view of the theorem, $H^2(\mathfrak{Q})$ is either *cyclic infinite* or of order 2. In the first case, $\mathfrak{Q}$ is said to be *orientable,* in the second, *non-orientable.*

13.4   THEOREM:

Given:

1°. *A monotone mapping $m:X \rightrightarrows X$, where X is a closed orientable 2-manifold.*

2°. *$\mathfrak{Q}$ is a hyper element of X.*

Conclusion: *$\mathfrak{Q}$ is orientable.*

Proof: By 6.8 there is a closed orientable 2-manifold Y, a suitable system $(\theta, \mathfrak{Y}, X)$, and a monotone mapping $\mu:Y \rightrightarrows \mathfrak{Y}$. Since $\mathfrak{Y}$ is a generalized cactoid, each true cyclic element of $\mathfrak{Y}$ is a closed orientable

2-manifold by 6.7.   There is a unique true cyclic element, $\mathfrak{E}$, of $\mathfrak{Y}$ such
that $\theta(\mathfrak{E}) = \mathfrak{L}$ (11.1)   Now by 13.3, $H^2(\mathfrak{L})$ is cyclic infinite.

__13.5__   THEOREM:

        Given:

                1°. *A monotone mapping $m: X \rightrightarrows \mathfrak{X}$, where X is a closed*
2-manifold.

                2°. *A simple region, $\mathfrak{U}$, in $\mathfrak{X}$.*

                3°. $U = m^{-1}(\mathfrak{U})$.

                4°. *A true cyclic element, $\mathfrak{L}$, of $\bar{\mathfrak{U}}$.*

                5°. U *is orientable.*

        Conclusion: $\mathfrak{L}$ *is orientable.*

Proof: The spaces and mappings of theorem 5.3 will be used without
further explanation.

        Since $\mathfrak{L}$ is a true cyclic element of $\bar{\mathfrak{U}}$, there is a unique mono-
tone retraction, $r: \bar{\mathfrak{U}} \rightrightarrows \mathfrak{L}$. If $\mathfrak{L}$ is not a 2-cell, it must be a hyper element
of $\mathfrak{X}$.   Hence $\mathfrak{L} \cap \dot{\mathfrak{U}} = 0$, by 12.2.   Therefore, $m^{-1}(\mathfrak{L}) \cap \dot{U} = 0$ and
$\phi m^{-1}(\mathfrak{L}) \cap \phi(\dot{U}) = 0$. Consequently, $rm(K_k) = x_k$, a single point on $\mathfrak{L}$, namely
the frontier point of the component of $\bar{\mathfrak{U}} - \mathfrak{L}$ containing $\mathfrak{R}_k$, $k = 1, \cdots, r$.
(See theorem 12.3.)

        Now consider the monotone mapping $\mu: \mathfrak{M} \rightrightarrows \mathfrak{L}$ provided by the formula
$\mu = rm\phi^{-1}$. The closed 2-manifold $\mathfrak{M}$ is orientable, hence $\mathfrak{L}$ is the monotone
image of an orientable 2-manifold. Moreover, it is quite clear that $\mathfrak{L}$, as a
mantoid, has a single hyper element, namely itself. Hence $H^2(\mathfrak{L})$ is cyclic infinite by 13.4.

__13.6__   If $\mathfrak{L}$ is a hyper element of a mantoid, then a set $\mathfrak{C}$ is said to
be a 2-cell *on* $\mathfrak{L}$ if and only if:

                1°. $\mathfrak{C}$ is a 2-cell.

                2°. $\mathfrak{C} \subset \mathfrak{L}$.

                3°. $\overline{\mathfrak{L} - \mathfrak{C}} \cap \mathfrak{C}$ is the Jordan curve bounding $\mathfrak{C}$.

        From the character of $\mathfrak{L}$ (see 11.1) it is clear that 2-cells
on $\mathfrak{L}$ exist, though a 2-cell *in* $\mathfrak{L}$, that is to say imbedded in $\mathfrak{L}$, need
not be a 2-cell *on* $\mathfrak{L}$. The bounding Jordan curve of such a 2-cell is
designated by $\dot{\mathfrak{C}}$ and $\mathfrak{C}^\circ$ is the notation for $\mathfrak{C} - \dot{\mathfrak{C}}$. Condition 3° states

that $\mathfrak{C}^\circ$ is open in $\mathfrak{L}$.

THEOREM:

Given:

1°. *A mantoid,* $\mathfrak{X}$.

2°. *A hyper element,* $\mathfrak{L}$, *of* $\mathfrak{X}$.

3°. *A 2-cell,* $\mathfrak{C}$, *on* $\mathfrak{L}$.

4°. $\mathfrak{L}$ *is orientable.*

Conclusion: $\mathfrak{k} : H^2(\mathfrak{C}^\circ) \approx H^2(\mathfrak{L})$.

Proof: Let $(\theta, \mathfrak{Y}, \mathfrak{X})$ be a suitable system and $\mathfrak{C}$ the true cyclic element of $\mathfrak{Y}$ such that $\theta(\mathfrak{C}) = \mathfrak{L}$ (see 11.1). Define $\bar{\theta} = \theta | \mathfrak{C}$, and let $x \in \mathfrak{P}^*$ if and only if $\bar{\theta}^{-1}(x)$ is non-degenerate. Define $\Omega^* = \bar{\theta}^{-1}(\mathfrak{P}^*)$. In view of 13.3 the closed manifold, $\mathfrak{C}$, is orientable. Since $\mathfrak{C}^\circ$ is open in $\mathfrak{L}$, clearly $\mathfrak{C}^\circ \cap \mathfrak{P}^* = 0$. If $\mathfrak{G} = \theta^{-1}(\mathfrak{C}^\circ)$ and $\underline{\theta} = \theta | \mathfrak{G}$, then $\underline{\theta} : \mathfrak{G} \approx \mathfrak{C}^\circ$. Now consider the commutative diagram

$$
\begin{array}{ccc}
H^2(\mathfrak{C}) & \xleftarrow{\quad a \quad} & H^2(\mathfrak{G}) \\[2pt]
\bar{\theta}^* \uparrow & & \uparrow \underline{\theta}^* \\[2pt]
H^2(\mathfrak{L}) & \xleftarrow{\quad \mathfrak{k} \quad} & H^2(\mathfrak{C}^\circ)
\end{array}
$$

The homomorphism $\underline{\theta}^*$ is an isomorphism onto by 8.5, a by 9.2, and $\bar{\theta}^*$ by 13.3; hence the theorem is true.

<u>13.7</u>   Suppose that $m : X \rightrightarrows \mathfrak{X}$ is a monotone mapping, where X is a closed 2-manifold. If $\mathfrak{U}$ is a *simple* region in $\mathfrak{X}$, and $\mathfrak{L}$ is a true cyclic element of $\bar{\mathfrak{U}}$, let $r : \bar{\mathfrak{U}} \rightrightarrows \mathfrak{L}$ be the monotone retraction. If $U = m^{-1}(\mathfrak{U})$ and $\underline{m} = m | U$, then $\underline{m} : U \rightrightarrows \mathfrak{U}$ is compact; hence $\underline{m}^* : H^2(\mathfrak{U}) \to H^2(U)$ has meaning by 8.2. Indeed, if $\bar{m} = m | \bar{U}$, then $\bar{m} : (\bar{U}, \dot{U}) \rightrightarrows (\bar{\mathfrak{U}}, \dot{\mathfrak{U}})$ and $\bar{m}$ is an extension of $\underline{m}$ to the compactifications $\bar{U}$ and $\bar{\mathfrak{U}}$ of U and $\mathfrak{U}$, respectively. Consequently, in view of the identification conventions in 8.1 and 8.2, one has the following diagram,

$$
\begin{array}{ccc}
H^2(\bar{\mathfrak{U}}, \dot{\mathfrak{U}}) & \xrightarrow{\quad\quad \bar{m}^* \quad\quad} & H^2(\bar{U}, \dot{U}) \\[2pt]
\| & & \| \\[2pt]
H^2(\mathfrak{U}) & \xrightarrow{\quad\quad \underline{m}^* \quad\quad} & H^2(U)
\end{array}
$$

and

(1)
$$\underline{m}^* = \bar{m}^*.$$

If $\underline{r} = r|\mathfrak{U}$, then $\underline{r}:U \rightrightarrows \mathfrak{L}^\circ$, where $\mathfrak{L}^\circ = \mathfrak{L}$ if $\mathfrak{L}$ is not a 2-cell, and if $\mathfrak{L}$ is a 2-cell then $\mathfrak{L}^\circ = \mathfrak{L} - \dot{\mathfrak{L}}$ as usual. In either case, let $\dot{\mathfrak{L}} = \mathfrak{L} - \mathfrak{L}^\circ$.

It is important to notice that $\underline{r}$ is not necessarily compact, hence $\underline{r}^*:H^2(\mathfrak{L}^\circ) \to H^2(\mathfrak{U})$ does not necessarily have meaning according to the conventions of 8.2. Nevertheless, there is a homomorphism from $H^2(\mathfrak{L}^\circ)$ into $H^2(\mathfrak{U})$ which is naturally determined by $\underline{r}$ though not induced by $\underline{r}$ in the ordinary sense (see 7.1 and 8.2).

To obtain this homomorphism notice that $\mathfrak{L}$ is a compactification of $\mathfrak{L}^\circ$ by the addition of $\dot{\mathfrak{L}}$, hence under the identification discussed in 8.1 it is possible to consider the following diagram:

$$
\begin{array}{ccccc}
H^2(\mathfrak{L},\dot{\mathfrak{L}}) & \xleftarrow{\quad i^* \quad} & H^2(\mathfrak{L},r(\dot{\mathfrak{U}})) & \xrightarrow{\quad r^* \quad} & H^2(\bar{\mathfrak{U}},\dot{\mathfrak{U}}) \\
\| & & \| & & \| \\
H^2(\mathfrak{L}^\circ) & \xleftarrow{\quad i \quad} & H^2(\mathfrak{L} - r(\dot{\mathfrak{U}})) & & H^2(\mathfrak{U})
\end{array}
$$

In view of 8.4 one has $i^* = i$, and hence by 8.6, both are isomorphisms onto. On defining

(2)
$$\underline{r}^* = r^*(i^*)^{-1} = r^* i^{-1}$$

one obtains a homomorphism

$$\underline{r}^*:H^2(\mathfrak{L}^\circ) \to H^2(\mathfrak{U}).$$

This is the homomorphism naturally determined by $\underline{r}$ mentioned above.

Observe that $\underline{r}^* = r^*$ if and only if i^* is the identity; that is, if and only if $r(\dot{\mathfrak{U}}) = \dot{\mathfrak{L}}$. But $r(\dot{\mathfrak{U}}) = \dot{\mathfrak{L}}$ if and only if $\underline{r}$ is compact. Consequently, if $\underline{r}$ is compact, then the homomorphism $\underline{r}^*:H^2(\mathfrak{L}^\circ) \to H^2(\mathfrak{U})$ induced by $\underline{r}$ in the ordinary way (see 8.1 and 8.2) agrees with the definition of $\underline{r}^*$ offered in (1) and is a further justification thereof.

If $\bar{\rho} = (rm)|\bar{U}$ and $\rho = (rm)|U$, then $\bar{\rho}:(\bar{U},\dot{U}) \rightrightarrows (\mathfrak{L},r(\dot{\mathfrak{U}}))$ and $\rho:U \rightrightarrows \mathfrak{L}^\circ = r(\mathfrak{U})$. Moreover

(3)
$$\text{(i)} \quad \bar{\rho} = r\bar{m}.$$
$$\text{(ii)} \quad \rho = \underline{r}\underline{m}.$$

The mapping ρ is not necessarily compact, but as in the case of

$\underline{r}$, there is a homomorphism $\rho^*:H^2(\mathfrak{L}^\circ) \to H^2(U)$ determined by it on employing the following diagram,

$$H^2(\mathfrak{L}, \, \dot{\mathfrak{L}}) \xleftarrow{\quad i^* \quad} H^2(\mathfrak{L}, \, r(\dot{\mathfrak{u}})) \xrightarrow{\quad \bar{\rho}^* \quad} H^2(\bar{U}, \, \dot{U})$$
$$\| \qquad\qquad\qquad \| \qquad\qquad\qquad \|$$
$$H^2(\mathfrak{L}^\circ) \xleftarrow{\quad i \quad} H^2(\mathfrak{L} - r(\dot{\mathfrak{u}})) \qquad\qquad H^2(U) \qquad\cdot$$

and defining

(4)
$$\rho^* = \bar{\rho}^*(i^*)^{-1} = \bar{\rho}^* i^{-1}.$$

As in the case for $\underline{r}$, if ρ is compact, then this definition of ρ^* yields the homomorphism induced by ρ as defined in 8.2.

It should be observed that these homomorphisms just defined behave in a proper fashion insofar as commutativity is concerned. For example, since $\rho = \underline{rm}$ by (3,ii) it would be advantageous to have $\rho^* = \underline{m}^*\underline{r}^*$. This is the case since,

$$\begin{aligned} \rho^* &= \bar{\rho}^* i^{-1} &&\text{by (4)} \\ &= \bar{m}^* r^* i^{-1} &&\text{by (3,i) and 7.3} \\ &= \bar{m}^* \underline{r}^* &&\text{by (2)} \\ &= \underline{m}^* \underline{r}^* &&\text{by (1).} \end{aligned}$$

It is highly important to note these conventions regarding $\underline{r}^*$ and ρ^*, and to bear in mind their satisfactory commutativity properties.

THEOREM: *With the above conventions, if U is orientable, then $\rho^*:H^2(\mathfrak{L}^\circ) \approx H^2(\mathfrak{u}).$*

Proof: In view of 13.2 there is a 2-cell, $\mathfrak{C}$, on $\mathfrak{L}$ such that $\mathfrak{C}^\circ \cap r(\mathfrak{u}) = 0$ and $m^{-1}r^{-1}(\mathfrak{C}^\circ)$ is an open 2-cell, G. If $\mu = \rho|G$, then $\mu:G \rightrightarrows \mathfrak{C}^\circ$ is a compact monotone mapping. Let $\bar{\mu} = \bar{\rho}$ and consider the following diagram:

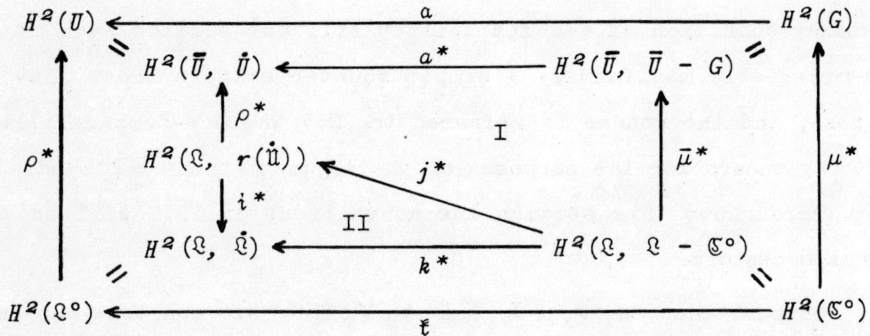

Commutativity holds in boxes I and II, consequently, in virtue of the definition of ρ^* and the conventions of 8.1, 8.2 and 8.4, commutativity holds on the perimeter of the diagram. But α is an isomorphism onto by 9.3 and ℓ by 9.3 if $\mathfrak{C}$ is a 2-cell and by 13.6 if $\mathfrak{C}$ is a hyper element of $\mathfrak{X}$. Finally μ^* is an isomorphism onto by 10.8. This proves the theorem.

Remark: Since $r(\mathfrak{U}) = \mathfrak{C}^\circ$, the conclusion of the theorem reads $\rho^*\!:\!H^2(r(\mathfrak{U})) \approx H^2(U)$. This form of the conclusion is often employed in the sequel.

14.   Compatibility

The F-criteria of this paper are phrased essentially in terms of the concepts to be introduced in this section. These concepts are *consistency* and *comparability* and are algebraic in character. (It is true that comparability is defined in point-set theoretic terms, but it could equally well have been defined in terms of 1-dimensional cohomology groups, thus rounding out the picture, since it will be observed that consistency is a 2-dimensional condition.) When both consistency and comparability hold, then the system under consideration will be said to be *compatible.*

As a further introduction to this section it should be stated that if one is dealing with orientable manifolds, then weaker conditions than comparability and consistency (to be known as *o*-comparability and *o*-consistency) may be employed. Comparability is a condition on all *normal* regions, *o*-comparability a condition on all *simple* regions. The reader may wonder if the weaker condition of *o*-comparability will not suffice even if dealing with non-orientable manifolds. A simple counter-example shows that this is *not* the case, and the reader is referred to 20.7 where *o*-comparability would not be strong enough for the purposes of the argument there employed.

Throughout this section the conventions of 13.7 will be employed on numerous occasions.

<u>14.1</u>   Suppose that $m_1\!:\!X_1 \rightrightarrows \mathfrak{X}$ and $m_2\!:\!X_2 \rightrightarrows \mathfrak{X}$ are monotone mappings from closed 2-manifolds, X_1 and X_2, while $\mathfrak{U}$ is a *simple* region in $\mathfrak{X}$ such that U_1 and U_2 are *orientable* where $U_i = m_i^{-1}(\mathfrak{U})$. Suppose $\{\mathfrak{C}_n\}$ is any

collection (finite or infinite) of true cyclic elements of $\bar{\mathfrak{U}}$ and $r_n{:}\bar{\mathfrak{U}} \rightrightarrows \mathfrak{L}_n$ is the monotone retraction, $n = 1, 2, 3, \cdots$. Define $\rho_{in} = (r_n m_i)\,|\,U_i$, $i = 1, 2$; $n = 1, 2, 3, \cdots$. By theorem 13.7, $\rho_{in}^*{:}H^2(r_n(\mathfrak{U})) \approx H^2(U_i)$, $i = 1, 2$; $n = 1, 2, 3, \cdots$.

The mappings m_1 and m_2 are said to be *consistent on* $\mathfrak{U}$ *with respect to* $\{\mathfrak{L}_n\}$ if and only if there is an isomorphism,

$$\eta{:}H^2(U_2) \approx H^2(U_1),$$

such that,

$$\rho_{1n}^* = \eta\rho_{2n}^*, \quad n = 1, 2, 3, \cdots.$$

The isomorphism η is called a *matching isomorphism*. Note that if m_1 and m_2 are consistent on $\mathfrak{U}$ with respect to $\{\mathfrak{L}_n\}$ and $\{\mathfrak{L}_n\}$ is not empty, then η is *unique*.

14.2   With the above conventions relative to m_1, m_2 and $\{\mathfrak{L}_n\}$, a collection $\{\mathfrak{C}_n\}$ of 2-cells is said to be *proper with respect to* m_1, m_2, $\{\mathfrak{L}_n\}$ if and only if $\mathfrak{C}_n$ is a 2-cell on $\mathfrak{L}_n$ and $m_i^{-1}r_n^{-1}(\mathfrak{C}_n^\circ) \equiv G_{in}$ is an open 2-cell, $i = 1, 2$; $n = 1, 2, 3, \cdots$.

In view of 13.2, it is a simple matter to show that a proper collection $\{\mathfrak{C}_n\}$ exists; in fact, select $\mathfrak{C}_n$ so that $\mathfrak{C}_n^\circ \subset [\mathfrak{L}_n - \{\mathfrak{F}(\mathfrak{L}_n^*, m_1) \cup \mathfrak{F}(\mathfrak{L}_n^*, m_2) \cup r_n(\mathring{\mathfrak{U}})\}]$, $n = 1, 2, 3, \cdots$.

Suppose $\mu_{in} = (r_n m_i)\,|\,G_{in}$, and consider the following diagram of isomorphisms onto for $i = 1, 2$; $n = 1, 2, 3, \cdots$:

$$H^2(U_i) \xleftarrow{\;a_{in}\;} H^2(G_{in}) \xleftarrow{\;\mu_{in}^*\;} H^2(\mathfrak{C}_n^\circ)$$

The mappings m_1 and m_2 are said to be *consistent on* $\mathfrak{U}$ *with respect to* $\{\mathfrak{C}_n\}$ if and only if there is an isomorphism $\eta{:}H^2(U_2) \approx H^2(U_1)$ such that $a_{1n}\mu_{1n}^* = \eta a_{2n}\mu_{2n}^*$, $n = 1, 2, 3, \cdots$. As above, η is called a matching isomorphism.

14.3   The equivalence of the two types of consistency is shown in the next theorem. In proofs, later in the paper, it is sometimes convenient to use one, at other times the other, definition. The proof follows at once from the last diagram of 13.7.

THEOREM:

Given:

$1\overset{\circ}{.}$ *Two monotone mappings, $m_1 : X_1 \rightrightarrows X$ and $m_2 : X_2 \rightrightarrows X$, where X_1 and X_2 are closed 2-manifolds.*

$2\overset{\circ}{.}$ *A simple region, $\mathfrak{U}$, in X such that U_1 and U_2 are orientable where $U_i = m_i^{-1}(\mathfrak{U})$, $i = 1, 2.$*

$3\overset{\circ}{.}$ *A finite or infinite collection, $\{\mathfrak{L}_n\}$, of true cyclic elements of $\overline{\mathfrak{U}}$.*

$4\overset{\circ}{.}$ *A collection of closed 2-cells, $\{\mathfrak{C}_n\}$, proper with respect to m_1, m_2, $\{\mathfrak{L}_n\}$.*

Conclusion: *The mappings m_1 and m_2 are consistent on $\mathfrak{U}$ with respect to $\{\mathfrak{L}_n\}$ if and only if they are consistent on $\mathfrak{U}$ with respect to $\{\mathfrak{C}_n\}$. Any matching isomorphism for one type of consistency is also a matching isomorphism for the other.*

<u>14.4</u>   The following result is useful in simplifying the concept of compatibility defined in 14.6.

THEOREM:

Given:

$1\overset{\circ}{.}$ *Two monotone mappings, $m_1 : X_1 \rightrightarrows X$ and $m_2 : X_2 \rightrightarrows X$, where X_1 and X_2 are closed 2-manifolds.*

$2\overset{\circ}{.}$ *If $\mathfrak{W}$ is a simple region in X such that W_i is orientable, where $W_i = m_i^{-1}(\mathfrak{W})$, $i = 1, 2,$ then m_1 and m_2 are consistent on $\mathfrak{W}$ with respect to every true cyclic element of $\overline{\mathfrak{W}}$ which is a 2-cell.*

$3\overset{\circ}{.}$ *$\mathfrak{U}$ is a simple region in X such that U_i is orientable where $U_i = m_i^{-1}(\mathfrak{U})$, $i = 1, 2.$*

Conclusion: *The mappings are consistent on $\mathfrak{U}$ with respect to every true cyclic element of $\overline{\mathfrak{U}}$.*

Proof: If there are no true cyclic elements in $\overline{\mathfrak{U}}$, then the consistency condition is vacuously fulfilled. If there is but one true cyclic element, $\mathfrak{L}$, in $\overline{\mathfrak{U}}$, then suppose $r : \overline{\mathfrak{U}} \rightrightarrows \mathfrak{L}$ is the monotone retraction. If $\rho_i = (rm_i)\,|\,U_i$, then $\rho_i : H^2(r(\mathfrak{U})) \approx H^2(U_i)$, $i = 1, 2,$ by 13.7. Hence there is certainly an isomorphism $\eta : H^2(U_2) \approx H^2(U_1)$ such that $\rho_1^* = \eta \rho_2^*.$

Now suppose that $\overline{\mathfrak{U}}$ has at least two true cyclic elements. Let $\mathfrak{L}_1$ be a fixed selection of a true cyclic element of $\overline{\mathfrak{U}}$, $r_1 : \overline{\mathfrak{U}} \rightrightarrows \mathfrak{L}_1$ the monotone retraction and $\rho_{i1} = (r_1 m_i)\,|\,U_i$, $i = 1, 2.$ In view of the preceding comments, there is certainly an isomorphism $\eta : H^2(U_2) \approx H^2(U_1)$ such that $\rho_{11}^* = \eta \rho_{21}^*$. Suppose that $\mathfrak{L}_2$ is any other true cyclic element of $\bar{\mathfrak{U}}$, while $r_2 : \bar{\mathfrak{U}} \rightrightarrows \mathfrak{L}_2$ is the monotone retraction and $\rho_{i2} = (r_2 m_i)|U_i$, $i = 1, 2$.

The theorem will be proved if it can be shown that $\rho_{12}^* = \eta \rho_{22}^*$.

There is an arc, a, in $\bar{\mathfrak{U}}$ with one end point, x_1, on $\mathfrak{L}_1$, the other end point, x_2, on $\mathfrak{L}_2$, and such that $a \cap (\mathfrak{L}_1 \cup \mathfrak{L}_2) = x_1 \cup x_2$. Recall that $\mathfrak{L}_i^*$, the unique hyper element of $\mathfrak{X}$ containing $\mathfrak{L}_i$, is obtained from a closed 2-manifold by a finite number of identifications, $i = 1, 2$; hence there is a finite number of 2-cells, $\mathfrak{C}_i$, $\cdots$, $\mathfrak{C}_{in_i}$, $i = 1, 2$, with the following properties:

(i) $\mathfrak{C}_{ij} \subset \mathfrak{L}_i^o$, $j = 1, \cdots, n_i$; $i = 1, 2$. (See 13.7 for $\mathfrak{L}_i^o$.)

(ii) $x_i \in \mathfrak{C}_{ij}^o$, $j = 1, \cdots, n_i$; $i = 1, 2$.

(iii) $\mathfrak{C}_{ij} \cap \mathfrak{C}_{ik} = x_i$, $j \neq k$; $j, k = 1, \cdots, n_i$; $i = 1, 2$.

(iv) $\overline{\mathfrak{L}_i - \bigcup \mathfrak{C}_{ij}} \cap \bigcup \mathfrak{C}_{ij}^o = 0$, $i = 1, 2$.

The set $\bigcup \dot{\mathfrak{C}}_{ij}$ cuts $\mathfrak{L}_i$ and x_i is in a component of $\mathfrak{L}_i - \bigcup \dot{\mathfrak{C}}_{ij}$, $i = 1, 2$. Hence there is a component, $\mathfrak{B}$, of $\mathfrak{U} - [\bigcup \dot{\mathfrak{C}}_{1j} \cup \bigcup \dot{\mathfrak{C}}_{2j}]$ containing x_1 and x_2. It follows that $\mathfrak{B}$ is a simple region in $\mathfrak{X}$, and $\mathfrak{C}_{11}, \cdots, \mathfrak{C}_{1n_1}, \mathfrak{C}_{21}, \cdots, \mathfrak{C}_{2n_2}$ are among the true cyclic elements of $\mathfrak{B}$.

Let $\mathfrak{C}_i = \mathfrak{C}_{i1}$ and suppose $s_i : \bar{\mathfrak{B}} \rightrightarrows \mathfrak{C}_i$ is the monotone retraction, $i = 1, 2$. Suppose $\mathfrak{C}_1$ and $\mathfrak{C}_2$ are 2-cells proper with respect to m_1, m_2, $\mathfrak{C}_1$, $\mathfrak{C}_2$ (see 14.2), while $m_i^{-1} s_j^{-1}(\mathfrak{C}_j^o) = G_{ij}$, $i, j = 1, 2$. Now G_{ij} is an open 2-cell; let $\nu_{ij} = (s_j m_i)|G_{ij}$ and $\mu_{ij} = (r_j m_i)|G_{ij}$, $i, j = 1, 2$. It follows that:

$$\nu_{ij} = \mu_{ij}, \quad i, j = 1, 2.$$

If $V_i = m_i^{-1}(\mathfrak{B})$, then $V_i \subset U_i$ and is orientable, $i = 1, 2$.

Now consider the following diagram for $i, j = 1, 2$:

$$H^2(V_i) \xleftarrow{\quad \mathfrak{b}_{ij} \quad} H^2(G_{ij}) \xleftarrow{\quad \nu_{ij}^* \quad} H^2(\mathfrak{C}_j^o)$$

By hypothesis, m_1 and m_2 are consistent on $\mathfrak{B}$ with respect to $\mathfrak{C}_1$ and $\mathfrak{C}_2$, and hence by 14.3 they are consistent on $\mathfrak{B}$ with respect to $\mathfrak{C}_1$ and $\mathfrak{C}_2$. Thus there is an isomorphism $\kappa : H^2(V_2) \approx H^2(V_1)$ such that $\mathfrak{b}_{1j} \nu_{1j}^* = \kappa \, \mathfrak{b}_{2j} \nu_{2j}^*$, $j = 1, 2$. But $\nu_{ij}^* = \mu_{ij}^*$, $i, j = 1, 2$, hence:

$$\mathfrak{b}_{1j} \mu_{1j}^* = \kappa \, \mathfrak{b}_{2j} \mu_{2j}^*, \quad j = 1, 2.$$

Now consider the following diagram for $j = 1, 2$:

$$
\begin{array}{ccccc}
H^2(V_1) & \xleftarrow{\quad\kappa\quad} & & & H^2(V_2) \\
\big\uparrow{\scriptstyle\mathfrak{b}_{1j}} & & \mathrm{I} & & \big\uparrow{\scriptstyle\mathfrak{b}_{2j}} \\
H^2(G_{1j}) & \xleftarrow{\;\mu^*_{1j}\;} & H^2(\mathbb{C}^\circ_j) & \xrightarrow{\;\mu^*_{2j}\;} & H^2(G_{2j}) \\
\big\downarrow{\scriptstyle a_{1j}} & & \mathrm{II} & & \big\downarrow{\scriptstyle a_{2j}} \\
H^2(U_1) & \xleftarrow[\quad\eta\quad]{} & & & H^2(U_2)
\end{array}
$$

It has been shown that commutativity holds in I for $j = 1, 2$, and by the selection of η, and 14.3, one has commutativity in II for $j = 1$. Consequently, the statement

$$ \eta = [a_{1j}\mathfrak{b}^{-1}_{1j}]\,\kappa\,[\mathfrak{b}_{2j}a^{-1}_{2j}] $$

holds for $j = 1$. But the isomorphisms in brackets are independent of j by 9.5. Hence the statement also holds for $j = 2$. Therefore,

$$ a_{12}\mu^*_{12} = \eta a_{22}\mu^*_{22}, $$

and by 14.3,

$$ \rho^*_{12} = \eta\rho^*_{22}. $$

14.5   The following readily established result shows that, roughly speaking, consistency is a hereditary property.

THEOREM:

Given:

1° *Two monotone mappings,* $m_1 : X_1 \rightrightarrows X$ *and* $m_2 : X_2 \rightrightarrows X$, *where* X_1 *and* X_2 *are closed 2-manifolds.*

2° m_1 *and* m_2 *are consistent on* $\mathfrak{U}$ *with respect to all the true cyclic elements of* $\bar{\mathfrak{U}}$.

3° $\mathfrak{V}$ *is a simple region of* X *in* $\mathfrak{U}$.

Conclusion: *The mappings are consistent on* $\mathfrak{V}$ *with respect to all the true cyclic elements of* $\bar{\mathfrak{V}}$.

14.6   *Definition:* Two monotone mappings, $m_1 : X_1 \rightrightarrows X$ and $m_2 : X_2 \rightrightarrows X$, where X_1 and X_2 are closed 2-manifolds, are said to be **consistent** if and only if for each *simple* region $\mathfrak{U}$ in X such that U_i is orientable, where

$U_i = m_i^{-1}(\mathfrak{U})$, $i = 1, 2,$ it is true that m_1 and m_2 are consistent on $\mathfrak{U}$
with respect to the *class of true cyclic elements of* $\overline{\mathfrak{U}}$ *which are 2-cells.*

In view of 14.4, an equivalent definition is obtained if the itali-
cized phrase above is replaced by *class of all true cyclic elements of* $\overline{\mathfrak{U}}$.

__14.7__   *Definition:* Two monotone mappings $m_1 : X_1 \rightrightarrows \mathfrak{X}$ and $m_2 : X_2 \rightrightarrows \mathfrak{X}$,
where X_1 and X_2 are closed 2-manifolds, are said to be *comparable* if and
only if for every *normal* region, $\mathfrak{U}$, in $\mathfrak{X}$ it is true that $U_1 \approx U_2$, where
$U_i = m_i^{-1}(\mathfrak{U})$, $i = 1, 2.$

Remark: Notice that if m_1 and m_2 are comparable, then $X_1 \approx X_2$
(let $\mathfrak{U} = \mathfrak{X}$), and further, U_1 is orientable if and only if U_2 is orientable.

__14.8__   *Definition:* Two monotone mappings, $m_1 : X_1 \rightrightarrows \mathfrak{X}$ and $m_2 : X_2 \rightrightarrows \mathfrak{X}$,
where X_1 and X_2 are closed 2-manifolds, are said to be *compatible* if and
only if they are both *comparable* and *consistent.*

__14.9__   It has been mentioned in chapter I that in spite of the fact that
consistency and comparability are sufficiently strong to provide an F-criterion,
regardless of whether one is considering mappings from 2-manifolds which are
orientable or not, it appears proper to consider also somewhat weaker conditions which are concerned only with orientable 2-manifolds. This is done here
in outline.

Suppose that $m : X \rightrightarrows \mathfrak{X}$ is a monotone mapping, where X is a
closed, *orientable* 2-manifold. If $\mathfrak{L}$ is a hyper element of $\mathfrak{X}$, then $\mathfrak{L}$ is
orientable by 13.4. Let $\mathfrak{T}$ be the true cyclic element of $\mathfrak{X}$ containing $\mathfrak{L}$,
and $r : \mathfrak{X} \rightrightarrows \mathfrak{T}$ the monotone retraction. Suppose that $\mathfrak{C}_i$ is a 2-cell on $\mathfrak{L}$
such that $m^{-1} r^{-1}(\mathfrak{C}_i^o)$ is an open 2-cell, G_i, in X, $i = 1, 2$ (see 13.1).
If $\mu_i = (rm)|G_i$,   then $\mu_i : G_i \rightrightarrows \mathfrak{C}_i^o$ is monotone, $i = 1, 2.$

Now *assume that* $\mathfrak{C}_1 \subset \mathfrak{C}_2$, and consider the following diagram:

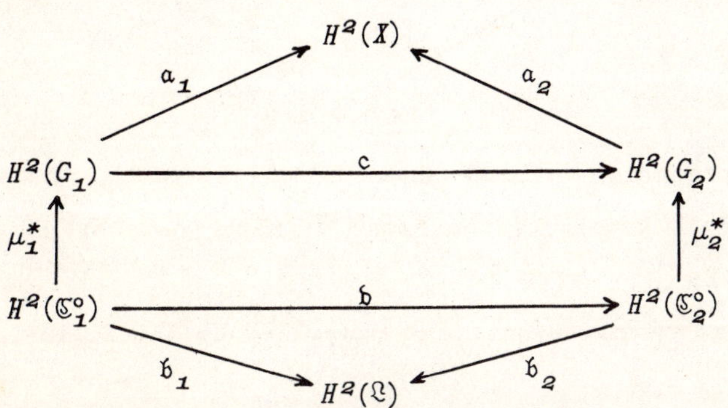

Commutativity holds in every box, and all the homomorphisms are isomorphisms onto. Hence:

$$a_1 \mu_1^* b_1^{-1} = a_2 \mu_2^* b_2^{-1}.$$

If $\mathfrak{C}_1 \not\subset \mathfrak{C}_2$, then there are 2-cells, $\mathfrak{C}_1^*$ and $\mathfrak{C}_2^*$, such that:

     (i) $\mathfrak{C}_i^* \subset \mathfrak{C}_i^\circ$, $i = 1, 2$.

     (ii) $\mathfrak{C}_1^* \cap \mathfrak{C}_2^* = 0$.

It is easy to see that there is a 2-cell $\mathfrak{C}$ whose interior is in $\mathfrak{L} - \mathfrak{F}(\mathfrak{L}, m)$ and such that $\mathfrak{C} \supset \mathfrak{C}_1^* \cup \mathfrak{C}_2^*$. These facts enable one to establish the above equality in general, that is to say, without the restriction $\mathfrak{C}_1 \subset \mathfrak{C}_2$.

If $\boldsymbol{x}$ is a generator for $H^2(X)$, then $b_1 (\mu_1^*)^{-1} a_1^{-1} (\boldsymbol{x})$ is a generator, $\mathfrak{l}$, for $H^2(\mathfrak{L})$ which is independent of the selection $\mathfrak{C}_1$. The generator, $\mathfrak{l}$, will be said to *determine* an orientation on $\mathfrak{L}$, and this orientation is said to be *induced* by the mapping, m, and the orientation on X determined by $\boldsymbol{x}$.

Definition: If $m_1 : X_1 \rightrightarrows X$ and $m_2 : X_2 \rightrightarrows X$ are monotone mappings from closed 2-manifolds, then m_1 and m_2 are said to be *o-consistent* if and only if X_1 and X_2 are *orientable* and there is an isomorphism, $\eta : H^2(X_2) \approx H^2(X_1)$ such that: if $\mathfrak{L}$ is any hyper element of X, $\mathfrak{T}$ is the true cyclic element of X containing $\mathfrak{L}$, $r : X \rightrightarrows \mathfrak{T}$ is the monotone retraction, $\mathfrak{C}$ is any 2-cell on $\mathfrak{L}$ such that $m^{-1} r^{-1}(\mathfrak{C}^\circ)$ is an open 2-cell, G_i, in X_i, while $\mu_i = (r m_i) | G_i$, $i = 1, 2$, then commutativity holds in the following diagram of isomorphisms onto:

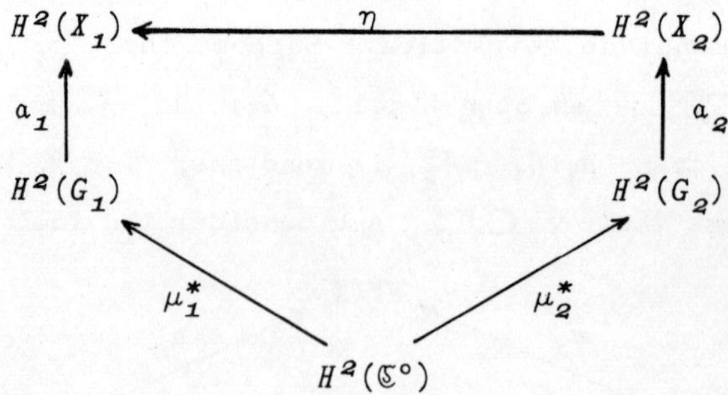

The preceding remarks show that it is sufficient to have commutativity for a single 2-cell $\check{\mathfrak{C}}$ on each $\mathfrak{L}$ with the property that $m_i^{-1} r^{-1}(\mathfrak{C}^\circ)$ is an open 2-cell.

An equivalent definition is supplied by the following simpler statement: if $m_1 : X_1 \rightrightarrows \mathfrak{X}$ and $m_2 : X_2 \rightrightarrows \mathfrak{X}$ are monotone mappings from closed 2-mani-folds, then m_1 and m_2 are said to be *o-consistent* if and only if X_1 and X_2 are *orientable* and there are fixed orientations on X_1 and X_2 such that for every hyper element, $\mathfrak{L}$, of $\mathfrak{X}$, the orientation induced on $\mathfrak{L}$, by the mapping m_i and the fixed orientation on X_i, is independent of $i = 1, 2$.

Definition: If $m_1 : X_1 \rightrightarrows \mathfrak{X}$ and $m_2 : X_2 \rightrightarrows \mathfrak{X}$ are monotone mappings from closed 2-manifolds, then they are said to be *o-comparable* if and only if X_1 and X_2 are *orientable*, and for any *simple* region, $\mathfrak{U}$, in $\mathfrak{X}$ it is true that $U_1 \approx U_2$ where $U_i = m_i(\mathfrak{U})$, $i = 1, 2$.

Remark: Notice that o-comparability is a condition on *simple*, not *normal*, regions.

Definition: Two monotone mappings, $m_1 : X_1 \rightrightarrows \mathfrak{X}$ and $m_2 : X_2 \rightrightarrows \mathfrak{X}$, where X_1 and X_2 are closed orientable 2-manifolds, are said to be *o-compatible* if and only if they are both o-comparable and o-consistent.

<u>14.10</u>   This concluding section is concerned with the special situation which arises in the event the monotone mappings are from 2-spheres; thus it complements the opening remarks of the chapter.

Suppose, therefore, that $m_1 : X_1 \rightrightarrows \mathfrak{X}$ and $m_2 : X_2 \rightrightarrows \mathfrak{X}$ are monotone mappings, where X_1 and X_2 are 2-spheres. In this case $\mathfrak{X}$ is a cactoid, that is, each true cyclic element of $\mathfrak{X}$ is a 2-sphere; consequently, the concepts of hyper element and true cyclic element coincide. And now it is easy to see that one has the following statement (cf. Youngs [17]):

THEOREM: *If $m_1 : X_1 \rightrightarrows \mathfrak{X}$ and $m_2 : X_2 \rightrightarrows \mathfrak{X}$ are monotone mappings, where X_1 and X_2 are 2-spheres, then m_1 and m_2 are o-consistent if and only if there is an isomorphism $\eta : H^2(X_2) \approx H^2(X_1)$ such that, if $\mathfrak{C}$ is any true cyclic element of $\mathfrak{X}$ and $r : \mathfrak{X} \rightrightarrows \mathfrak{C}$ the monotone retraction, then one has commutativity in the diagram below:*

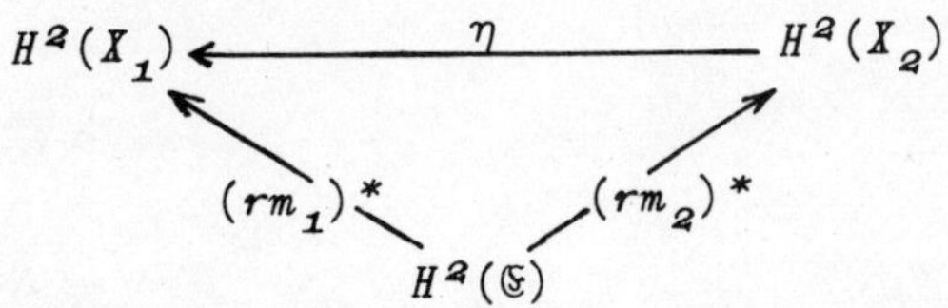

Remark: The theorem is true if X_1 and X_2 are orientable

2—manifolds and X is a generalized cactoid, for in this event each hyper element of X is a true cyclic element of X, and conversely.

The next theorem shows that the condition of comparability (and hence o—comparability) is automatically fulfilled.

THEOREM: *If $m_1 : X_1 \rightrightarrows X$ and $m_2 : X_2 \rightrightarrows X$ are monotone mappings from 2-spheres X_1 and X_2, then m_1 and m_2 are comparable.*

Proof: Suppose $\mathfrak{U}$ is any normal region in X while $\mathring{\mathfrak{U}}$ has r components. Let $U_i = m_i^{-1}(\mathfrak{U})$ and suppose M_i is a 2-manifold associated with U_i, $i = 1, 2$. Since $U_i \approx M_i^o$, $i = 1, 2$, it is sufficient to show that $M_1 \approx M_2$. But $\mathring{M}_i$ has r components and is a subset of the 2—sphere X_i, $i = 1, 2$, and this is sufficient to guarantee that $M_1 \approx M_2$.

CHAPTER IV

THE NECESSITY ARGUMENT

The data collected in the earlier chapters now enable one to approach the representation problem itself, and in conformity with the title to this chapter the discussion is directed towards showing that certain conditions are necessary for two mappings to be Fréchet equivalent.

It must be remembered that one is dealing with two general classes of 2—manifolds, namely closed 2—manifolds and 2—manifolds with boundary. (In addition, of course, the 2—manifolds may be either orientable or non—orientable).

15. The case for 2—manifolds with boundary

It was stated in 2.13 that the discussion for mappings from 2—manifolds with boundary would be reduced to that for mappings from closed 2—manifolds. The manner in which this is done for the necessity argument is shown in this section culminating in theorem 15.7, which may be classified as a modified reduction theorem (see 1.8).

15.1 *Definition:* A 2—manifold, H, is said to be *withdrawn* from a closed 2—manifold, X, if and only if:

    1° $H \subset X$, but $H \neq X$.

    2° Each component of $X - H$ is an open 2—cell.

If H is a 2—manifold with boundary, suppose X is a 2—manifold obtained by capping the bounding curves of H with 2—cells. Then H is a 2—manifold withdrawn from X. Conversely, if H is a 2—manifold withdrawn from X, then X is a 2—manifold obtained by capping the bounding curves of H with 2—cells.

15.2 *Definition:* A mapping, m, is said to be *admissable* with respect to X, H, $\mathfrak{X}$ and $\mathfrak{H}$ if and only if:

    1° X is a *closed* 2—manifold.

2° $H = X$ or H is a 2-manifold withdrawn from X.

3° $m : X \rightrightarrows \mathfrak{X}$.

4° m is monotone.

5° $m(H) = \mathfrak{H}$.

6° If $x \in \overline{X - H}$, then $x = m^{-1}m(x)$.

A mapping, m, is said to be *weakly admissible* with respect to X, H, $\mathfrak{X}$ and $\mathfrak{H}$ if and only if the first five conditions above hold and 6° is weakened to read:

6° If $x \in X - H$, then $x = m^{-1}m(x)$.

Remark: Notice that if m is weakly admissible with respect to X, H, $\mathfrak{X}$ and $\mathfrak{H}$, but H is *not* withdrawn from X, then $H = X$, from which it follows directly that $\mathfrak{H} = \mathfrak{X}$ and m is admissible with respect to X, H, $\mathfrak{X}$ and $\mathfrak{H}$. In fact, m is simply a monotone mapping from X onto $\mathfrak{X}$. Conversely, if $m : X \rightrightarrows \mathfrak{X}$ is monotone, where X is a closed 2-manifold, then, setting $H = X$ and $\mathfrak{H} = \mathfrak{X}$, the mapping m is admissible with respect to X, H, $\mathfrak{X}$ and $\mathfrak{H}$.

<u>15.3</u>   The following result is easily established.

THEOREM:

Given, for $i = 1$, 2:

1° m_i *is weakly admissible with respect to* X_i, H_i, $\mathfrak{X}$ *and* $\mathfrak{H}$.

2° $\mathfrak{R}^1$, $\mathfrak{R}^2$, $\mathfrak{R}^3$, $\cdots$, *are the components of* $\mathfrak{X} - \mathfrak{H}$.

Conclusion, *for* $i = 1$, 2:

3° $R_i^k \equiv m_i^{-1}(\mathfrak{R}^k)$, $k = 1$, 2, 3, $\cdots$, *are the components of* $X_i - H_i$.

4° $m_i(\mathring{R}_i^k) = \mathring{\mathfrak{R}}^k$, $k = 1$, 2, 3, $\cdots$.

Immediate consequences:

5° *There are but a finite number of components to* $\mathfrak{X} - \mathfrak{H}$.

6° $X_1 - H_1$ *and* $X_2 - H_2$ *have the same number of components.*

Remark: In the sequel it will be understood that if m_1 and m_2 are weakly admissible with respect to X_i, H_i, $\mathfrak{X}$ and $\mathfrak{H}$, while the components of $X_i - H_i$, if any, are enumerated as R_i^1, $\cdots$, R_i^n, $i = 1$, 2, then $m_1(R_1^k) = \mathfrak{R}^k = m_2(R_2^k)$, $k = 1$, $\cdots$, n. The components R_1^k and R_2^k are known as *corresponding* components of $X_1 - H_1$ and $X_2 - H_2$, respectively, $k = 1$, $\cdots$, n.

15.4   The lemmas which follow, together with the theorem of 10.1 consti-
tute the key to this section.

Preliminary remark: Suppose Θ is the totality of real numbers and
$\eta: \Theta \approx \Theta$: (Clearly η is either a strictly increasing or a strictly decreasing
continuous function.)

For $0 \leqq r \leqq 1$ and $\theta \in \Theta$, define

$$(1) \qquad \phi_r(\theta) = \begin{cases} (1 - r)\theta + r\eta(\theta) & \text{if } \eta \text{ is increasing.} \\[2mm] -(1 - r)\theta + r\eta(\theta) & \text{if } \eta \text{ is decreasing.} \end{cases}$$

Notice that

$$(2) \qquad \phi_r: \Theta \approx \Theta, \quad 0 \leqq r \leqq 1.$$

If

$$(3) \qquad \eta_{r1} = \phi_r^{-1}, \quad 0 \leqq r \leqq 1.$$

and

$$(4) \qquad \eta_{r2} = \eta\eta_{r1}, \quad 0 \leqq r \leqq 1,$$

then

$$(5) \qquad \eta_{ri}: \Theta \approx \Theta, \quad 0 \leqq r \leqq 1, \quad i = 1, 2.$$

If $\bar{\theta}_1 > \bar{\theta}_2$, and η is increasing, then

$$\phi_r(\bar{\theta}_1) - \phi_r(\bar{\theta}_2) = (1 - r)(\bar{\theta}_1 - \bar{\theta}_2) + r[\eta(\bar{\theta}_1) - \eta(\bar{\theta}_2)],$$

while if η is decreasing, then

$$\phi_r(\bar{\theta}_2) - \phi_r(\bar{\theta}_1) = (1 - r)(\bar{\theta}_1 - \bar{\theta}_2) + r[\eta(\bar{\theta}_2) - \eta(\bar{\theta}_1)].$$

Consequently, if $\bar{\theta}_1 \in \Theta \ni \bar{\theta}_2$, $0 \leqq r \leqq 1$, then

$$|\phi_r(\bar{\theta}_1) - \phi_r(\bar{\theta}_2)| = (1 - r)|\bar{\theta}_1 - \bar{\theta}_2| + r|\eta(\bar{\theta}_1) - \eta(\bar{\theta}_2)|,$$

$$(6) \qquad (1 - r)|\bar{\theta}_1 - \bar{\theta}_2| \leqq |\phi_r(\bar{\theta}_1) - \phi_r(\bar{\theta}_2)| \geqq r|\eta(\bar{\theta}_1) - \eta(\bar{\theta}_2)|.$$

For $\theta_1 \in \Theta \ni \theta_2$, let $\bar{\theta}_i = \phi_r^{-1}(\theta_i)$, $i = 1, 2$.

Then (3), (4) and (6) provide that

$$(1 - r)|\eta_{r1}(\theta_1) - \eta_{r1}(\theta_2)| \leqq |\theta_1 - \theta_2| \geqq r|\eta_{r2}(\theta_1) - \eta_{r2}(\theta_2)|.$$

Hence $\theta_1 \in \Theta \ni \theta_2$, $0 < r < 1$ implies

$$(\text{i}) \qquad |\eta_{r1}(\theta_1) - \eta_{r1}(\theta_2)| \leqq [1/(1 - r)]|\theta_1 - \theta_2|,$$

$$(7)$$

$$(\text{ii}) \qquad |\eta_{r2}(\theta_1) - \eta_{r2}(\theta_2)| \leqq [1/r]|\theta_1 - \theta_2|.$$

Similar elementary calculations show that:

If $\eta(\theta + 2\pi) = \eta(\theta) + 2\pi$ and $|\eta(\theta) - \theta| < 2\pi$, $\theta \in \Theta$, then for $0 \leqq r \leqq 1$:

$\qquad$ (i) $\qquad\qquad \phi_r(\theta + 2\pi) = \phi_r(\theta) + 2\pi$ and $|\phi_r(\theta) - \theta| < 2\pi r$, $\qquad \theta \in \Theta$,

(8) $\quad$ (ii) $\qquad\qquad \eta_{r_1}(\theta + 2\pi) = \eta_{r_1}(\theta) + 2\pi$ and $|\eta_{r_1}(\theta) - \theta| < 2\pi r$, $\quad \theta \in \Theta$,

$\qquad$ (iii) $\qquad\qquad \eta_{r_2}(\theta + 2\pi) = \eta_{r_2}(\theta) + 2\pi$, $\qquad\qquad\qquad\qquad\qquad \theta \in \Theta$.

If $\eta(\theta + 2\pi) = \eta(\theta) - 2\pi$ and $|\eta(\theta) + \theta| < 2\pi$, $\theta \in \Theta$, then for $0 \leqq r \leqq 1$:

$\qquad$ (i) $\qquad\qquad \phi_r(\theta + 2\pi) = \phi_r(\theta) - 2\pi$ and $|\phi_r(\theta) + \theta| < 2\pi r$, $\qquad \theta \in \Theta$,

(9) $\quad$ (ii) $\qquad\qquad \eta_{r_1}(\theta + 2\pi) = \eta_{r_1}(\theta) - 2\pi$ and $|\eta_{r_1}(\theta) + \theta| < 2\pi r$, $\quad \theta \in \Theta$,

$\qquad$ (iii) $\qquad\qquad \eta_{r_2}(\theta + 2\pi) = \eta_{r_2}(\theta) + 2\pi$, $\qquad\qquad\qquad\qquad\qquad \theta \in \Theta$.

Moreover, under each of the hypotheses above, one has:

$\qquad$ (i) $\qquad\qquad |\phi_{r_1}(\theta_1) - \phi_{r_2}(\theta_2)| \leqq |\theta_1 - \theta_2| + |\eta(\theta_1) - \eta(\theta_2)| + 2\pi|r_1 - r_2|,$

(10)

$\qquad$ (ii) $\qquad\qquad L|\eta_{r_1 1}(\theta_1) - \eta_{r_2 1}(\theta_2)| \leqq |\theta_1 - \theta_2| + 2\pi|r_1 - r_2|.$

for $\theta \in \Theta$, $0 \leqq r_i \leqq 1$, $i = 1, 2$, where $L = max\ [(1 - r_1), (1 - r_2)]$.

$\qquad$ LEMMA:

$\qquad$ Given:

$\qquad\qquad$ 1$\overset{\circ}{.}$ *Three 2-cells C, C_1 and C_2.*

$\qquad\qquad$ 2$\overset{\circ}{.}$ *A sequence $\{h_n\}$ of homeomorphisms $h_n: C_1 \approx C_2$, $n = 1, 2,$*

3, $\cdots$.

$\qquad\qquad$ Conclusion: *There is a sequence $\{h_{in}\}$ of homeomorphisms $h_{in}:$*
$C \approx C_i$, $n = 1, 2, 3, \cdots$, $i = 1, 2$, such that for $i = 1, 2$:

$\qquad\qquad$ 3$\overset{\circ}{.}$ *$\{h_{in}\}$ is equicontinuous.*

$\qquad\qquad$ 4$\overset{\circ}{.}$ *For any $\epsilon > 0$ there is a $\delta > 0$ such that $\rho\{x, \dot{C}\} \geqq \epsilon$*
implies $\rho\{h_{in}(x), \dot{C}_i\} \geqq \delta$, $n = 1, 2, 3, \cdots$.

$\qquad\qquad$ 5$\overset{\circ}{.}$ *If K_i is a compact subset of $\overset{\circ}{C}_i$, then $\{h_{in}^{-1}|K_i\}$ is*
equicontinuous.

$\qquad\qquad$ 6$\overset{\circ}{.}$ *If $\dot{h}_{in} = h_{in}|\dot{C}$, $i = 1, 2$, then $\dot{h}_n = \dot{h}_{2n}\dot{h}_{1n}^{-1}$.*

$\qquad$ *Proof:* The lemma will certainly follow if it is true in case
$C = \{z|\ |z| \leqq 1\}$ in a complex z-plane, and $C_1 = C = C_2$.

$\qquad$ Using the notation of the preliminary remark, for $n = 1, 2, 3, \cdots$
it is possible to select a homeomorphism $\eta^n: \Theta \approx \Theta$ such that:

$$h_n(1,\ \theta) = (1,\ \eta^n(\theta)), \qquad \theta \in \Theta,$$

and

$$|\eta^n(\theta) - \theta| < 2\pi, \qquad \theta \in \mathit{\Theta},$$

or

$$|\eta^n(\theta) + \theta| < 2\pi, \qquad \theta \in \mathit{\Theta},$$

It is clear that one of the following statements holds:

(11)
$$\eta^n(\theta + 2\pi) = \eta^n(\theta) + 2\pi, \qquad \theta \in \mathit{\Theta},$$
$$\eta^n(\theta + 2\pi) = \eta^n(\theta) - 2\pi, \qquad \theta \in \mathit{\Theta}.$$

Apply the preliminary remark with η replaced by η^n and $r = \frac{1}{2}$ to define, for $n = 1, 2, 3, \cdots, i = 1, 2, \ -\infty < \theta < \infty$,

$$\dot{h}_{in}(1, \ \theta) = (1, \ \eta^n_{\frac{1}{2}i}(\theta)),$$

where $\eta^n_{\frac{1}{2}i}$ is defined as in (3) and (4) for $i = 1, 2$.

Notice that for $n = 1, 2, 3, \cdots, \ i = 1, 2,$

(12)
$$\dot{h}_{in} : \dot{C} \approx \dot{C}. \quad (\text{See } (5), (11), (8) \text{ and } (9).)$$
$$\dot{h}_n = \dot{h}_{2n}\dot{h}_{1n}^{-1}. \quad (\text{See } (4).)$$

For notational convenience, let

$$\eta^n_i = \eta^n_{\frac{1}{2}i}, \quad i = 1, 2.$$

Now use the preliminary remark again with η replaced by η^n_i to define, for $n = 1, 2, 3, \cdots, \ i = 1, 2, \ -\infty < \theta < \infty, \ 0 \leqq r \leqq 1,$

(13)
$$h_{in}(r, \ \theta) = (r, \ \phi^n_{ir}(\theta)),$$

where ϕ^n_{ir} is defined as in (1).

Notice that for $n = 1, 2, 3, \cdots, \ i = 1, 2, \ -\infty < \theta < \infty,$

(14)
$$h_{in}(1, \ \theta) = (1, \ \eta^n_i(\theta)) = (1, \ \eta^n_{\frac{1}{2}i}(\theta)) = \dot{h}_{in}(1, \ \theta)$$

Moreover:

$$h_{in} : C \approx C. \quad (\text{See } (13), (2), (11), (8) \text{ and } (9).)$$

$$\{h_{in}\} \text{ is equicontinuous.} \quad (\text{See } (10,i), \text{ the definition}$$

of $\{\eta^n_i\}$ and (7).)

$$\text{If } |z| \leqq r < 1, \quad \text{then} \quad |h_{in}(z)| \leqq r. \quad (\text{See } (13).)$$

$$\text{If } K_r = \{z \mid |z| \leqq r < 1\}, \quad \text{then} \quad \{h_{in}^{-1}|K_r\} \text{ is equicon-}$$

tinuous. (See (13), (3) and (10),ii).)

$$\dot{h}_n = \dot{h}_{2n}\dot{h}_{1n}^{-1}. \quad (\text{See } (14) \text{ and } (12).)$$

Consequently, $\{h_{in}\}$ has the desired properties, $i = 1, 2.$

<u>15.5</u>   LEMMA:

Given:

1° *A 2-cell,* $C_i, \ i = 1, 2.$

$2^{\circ}.$ *A mapping,* $g_i:C_i \rightrightarrows \mathbb{C}_i,$ *such that if* $x_i \in C_i^{\circ},$ *then* $x_i = g_i^{-1}g_i(x_i),$ $i = 1, 2.$

$3^{\circ}.$ *A sequence,* $\{\dot{h}_n\},$ *of homeomorphisms,* $\dot{h}_n:\dot{C}_1 \approx \dot{C}_2,$ $n = 1, 2, 3, \cdots,$ *and a homeomorphism,* $\dot{\mathfrak{h}}:\dot{\mathbb{C}}_1 \approx \dot{\mathbb{C}}_2,$ *where* $\dot{\mathbb{C}}_i = g_i(\dot{C}_i),$ $i = 1, 2,$ *such that if* $\dot{g}_i = g_i|\dot{C}_i,$ $i = 1, 2,$ *then:*

$$\dot{g}_2\dot{h}_n \rightrightarrows \dot{\mathfrak{h}}\dot{g}_1.$$

Conclusion: *There is:*

$4^{\circ}.$ *A homeomorphic extension,* $\mathfrak{h}:\mathbb{C}_1 \approx \mathbb{C}_2,$ *of* $\dot{\mathfrak{h}}.$

$5^{\circ}.$ *A subsequence,* $\{\dot{h}_{n_j}\},$ *of* $\{\dot{h}_n\},$ *and a homeomorphic extension,* $h_{n_j}:C_1 \approx C_2,$ *of* $\dot{h}_{n_j},$ $j = 1, 2, 3, \cdots,$ *such that:*

$$g_2 h_{n_j} \rightrightarrows \mathfrak{h}g_1.$$

Proof: Let $C = \{z \mid |z| \leq 1\}$ in a complex z-plane. Using the preceding lemma, one has a sequence $\{h_{in}\}$ of homeomorphisms $h_{in}:C \approx C_i,$ $n = 1, 2, 3, \cdots,$ $i = 1, 2,$ such that properties 3°, 4°, 5° and 6° of the conclusion are satisfied.

On defining

$$(1) \qquad h_n = h_{2n}h_{1n}^{-1}, \quad n = 1, 2, 3, \cdots,$$

one has

$$(2) \qquad h_n:C_1 = C_2, \quad n = 1, 2, 3, \cdots,$$

and by item 6° of the preceding lemma, $x \in \dot{C}_1$ implies

$$h_n(x) = h_{2n}h_{1n}^{-1}(x) = \dot{h}_{2n}\dot{h}_{1n}^{-1}(x) = \dot{h}_n(x), \quad n = 1, 2, 3, \cdots.$$

Hence

$$(3) \qquad h_n \text{ is an extension of } \dot{h}_n, \quad n = 1, 2, 3, \cdots.$$

In view of item 3° of the preceding lemma, a classical result guarantees that there is a sequence of integers

$$0 < n_1 < n_2 < n_3 < \cdots$$

such that $\{h_{in_j}\}$ converges uniformly for $i = 1, 2.$ For notational convenience, from this point on let $\{h_{in}\}$ designate the uniformly convergent subsequence, and s_i the limit mapping, $i = 1, 2.$ Hence

$$(4) \qquad h_{in} \rightrightarrows s_i \text{ and } s_i:C \rightrightarrows C_i, \quad i = 1, 2.$$

The mapping s_i is to be examined in some detail, $i = 1, 2.$

For fixed $i = 1, 2,$ if $x \in C^{\circ}$ and $\bar{x} \in s_i^{-1}s_i(x),$ then $h_{in}(x) \to s_i(x)$ and $h_{in}(\bar{x}) \to s_i(\bar{x}) = s_i(x).$ In view of item 4° of the preceding lemma, $\quad s_i(x) \in C_i^o.\quad$ Hence if

$$K_i = s_i(x) \cup \bigcup h_{in}(x) \cup \bigcup h_{in}(\bar{x}),$$

then $\ K_i\ $ is a compact set in $\ C_i^o.\ $ Therefore, by item $5°$ of the preceding lemma, there is a uniformly convergent subsequence $\{h_{in_k}^{-1}|K_i\}\ $ of $\ \{h_{in}^{-1}|K_i\}.$

    The sequence

$$h_{in_1}(x),\ h_{in_2}(\bar{x}),\ h_{in_3}(x),\ h_{in_4}(\bar{x}),\ \cdots$$

lies in $\ K_i\ $ and converges to $\ s_i(x).\ $ Hence uniform convergence of $\ \{h_{in_k}^{-1}|K_i\}$ implies that the sequence

$$x,\ \bar{x},\ x,\ \bar{x},\ \cdots$$

converges. This shows that $\ \bar{x} = x\ $ and

$$(5) \qquad x = s_i^{-1} s_i(x), \quad x \in C^o, \quad i = 1,\ 2.$$

    Consequently, if $\ \dot{s}_i = s_i|\dot{C},\ $ then

$$(6) \qquad \dot{s}_i : \dot{C} \rightrightarrows \dot{C}_i \quad \text{and} \quad \dot{s}_i^{-1}(x_i) = s_i^{-1}(x_i), \quad x_i \in \dot{C}_i, \quad i = 1,\ 2.$$

    Notice that

$$\bar{\rho}\{\dot{g}_2\dot{h}_n\dot{h}_{1n},\ \dot{\mathfrak{h}}\dot{g}_1\dot{s}_1\} \le \bar{\rho}\{\dot{g}_2\dot{h}_n\dot{h}_{1n},\ \dot{\mathfrak{h}}\dot{g}_1\dot{h}_{1n}\} + \bar{\rho}\{\dot{\mathfrak{h}}\dot{g}_1\dot{h}_{1n},\ \dot{\mathfrak{h}}\dot{g}_1\dot{s}_1\}$$
$$\le \bar{\rho}\{\dot{g}_2\dot{h}_n,\ \dot{\mathfrak{h}}\dot{g}_1\} + \bar{\rho}\{\dot{\mathfrak{h}}\dot{g}_1\dot{h}_{1n},\ \dot{\mathfrak{h}}\dot{g}_1\dot{s}_1\}$$

But $\ \dot{g}_2\dot{h}_n \rightrightarrows \dot{\mathfrak{h}}\dot{g}_1\ $ by hypothesis, and $\ \dot{h}_{1n} \rightrightarrows \dot{s}_1\ $ by (4), hence $\ \dot{\mathfrak{h}}\dot{g}_1\dot{h}_{1n} \rightrightarrows \dot{\mathfrak{h}}\dot{g}_1\dot{s}_1,$ and

$$\dot{g}_2\dot{h}_n\dot{h}_{1n} \rightrightarrows \dot{\mathfrak{h}}\dot{g}_1\dot{s}_1.$$

But by (1) and (4),

$$\dot{g}_2\dot{h}_n\dot{h}_{1n} = \dot{g}_2\dot{h}_{2n} \rightrightarrows \dot{g}_2\dot{s}_2.$$

Hence

$$\dot{g}_2\dot{s}_2 = \dot{\mathfrak{h}}\dot{g}_1\dot{s}_1.$$

Using item $2°$ of the hypothesis and (6), this implies that,

$$(7) \qquad \dot{\mathfrak{h}}(\mathfrak{x}_1) = g_2 s_2 s_1^{-1} g_1^{-1}(\mathfrak{x}_1), \qquad \mathfrak{x}_1 \in \dot{\mathfrak{C}}_1.$$
$$\dot{\mathfrak{h}}^{-1}(\mathfrak{x}_2) = g_1 s_1 s_2^{-1} g_2^{-1}(\mathfrak{x}_2), \qquad \mathfrak{x}_2 \in \dot{\mathfrak{C}}_2.$$

    Now define

$$(8) \qquad \mathfrak{h}(\mathfrak{x}_1) = g_2 s_2 s_1^{-1} g_1^{-1}(\mathfrak{x}_1), \qquad \mathfrak{x}_1 \in \mathfrak{C}_1.$$

In view of (5) and (7) together with item $2°$ of the hypothesis this is a one–to–one transformation and

$$\mathfrak{h}|\dot{C}_1 = \dot{\mathfrak{h}}.$$

    To prove continuity, suppose that $\ \mathfrak{x}_1^n \to \mathfrak{x}_1\ $ in $\ \mathfrak{C}_1.\ $ Then

$$\lim \sup\ (g_1 s_1)^{-1}(\mathfrak{x}_1^n) \subset (g_1 s_1)^{-1}(\mathfrak{x}_1),$$

hence

$$\lim \sup \; g_2 s_2 s_1^{-1} g_1^{-1}(x_1^n) \subset \mathfrak{H}(x_1);$$

that is

$$h(x_1^n) \to h(x_1).$$

Therefore

(9) $$\mathfrak{H} : \mathbb{C}_1 \approx \mathbb{C}_2.$$

     In view of (2), (3) and (9) it remains but to show that $g_2 h_n \rightrightarrows \mathfrak{H} g_1$.
Observe that for $n = 1, 2, 3, \cdots$

$$\bar{\rho}\{g_2 h_n, \; \mathfrak{H} g_1\} = \bar{\rho}\{g_2 h_n h_{1n}, \; \mathfrak{H} g_1 h_{1n}\}$$
$$\leq \bar{\rho}\{g_2 h_n h_{1n}, \; g_2 s_2\} + \bar{\rho}\{g_2 s_2, \; \mathfrak{H} g_1 s_1\}$$
$$+ \bar{\rho}\{\mathfrak{H} g_1 s_1, \; \mathfrak{H} g_1 h_{1n}\}.$$

In view of (8) the second term vanishes. Moreover $g_2 h_n h_{1n} = g_2 h_{2n}$ by (1),
while $g_2 h_{2n} \rightrightarrows g_2 s_2$ and $\mathfrak{H} g_1 h_{1n} \rightrightarrows \mathfrak{H} g_1 s_1$ by (4). Hence

$$g_2 h_n \rightrightarrows \mathfrak{H} g_1$$

and the lemma is proved.

15.6   THEOREM:

     Given:

        1°. *A closed 2-manifold,* X_i, $i = 1, 2$.

        2°. *A 2-manifold,* H_i, *withdrawn from* X_i, $i = 1, 2$.

        3°. *A monotone mapping,* $\widetilde{m}_i : H_i \rightrightarrows \widetilde{\mathfrak{H}}$, $i = 1, 2$.

        4°. $\widetilde{m}_1 \sim \widetilde{m}_2$.

     Conclusion: *There is:*

        5°. *A mapping,* m_i, *weakly admissible with respect to* X_i,
H_i, $\mathfrak{X}$ *and* $\mathfrak{H}$, $i = 1, 2$, *such that* $m_1 \sim m_2$.

        6°. *A homeomorphism,* $h : \mathfrak{H} \approx \widetilde{\mathfrak{H}}$, *such that* $\widetilde{m}_i = h(m_i | H_i)$,
$i = 1, 2$.

     Proof: By theorem 3.5 it is known that:

     There is a monotone mapping, $\mu_i : X_i \rightrightarrows \mathfrak{X}_i$, such that for $i = 1, 2$:

        (i)   If $\mu_i(H_i) = \mathfrak{H}_i$, then $\mu_i^{-1}(\mathfrak{H}_i) = H_i$,

(1)

        (ii)   If $x_i \in X_i - H_i$, then $x_i = \mu_i^{-1} \mu_i(x_i)$,

and a light mapping, $\lambda_i : \mathfrak{H}_i \rightrightarrows \widetilde{\mathfrak{H}}$, such that if $\underline{\mu}_i = \mu_i | H_i$, $i = 1, 2$, then:

(2) $$\widetilde{m}_i = \lambda_i \underline{\mu}_i.$$

Since $\widetilde{m}_i$ is itself monotone, the mapping

(3) $\qquad\qquad\qquad \lambda_i$ is a homeomorphism, $\quad i = 1, 2.$

Since

$$\widetilde{m}_1 \sim \widetilde{m}_2,$$

there is a sequence, $\{\widetilde{h}_n\}$, of homeomorphisms,

$$\widetilde{h}_n : H_1 \approx H_2, \quad n = 1, 2, 3, \cdots,$$

such that

(4) $\qquad\qquad\qquad \widetilde{m}_2 \widetilde{h}_n \rightrightarrows \widetilde{m}_1.$

Hence $\lambda_2(\underline{\mu}_2 \widetilde{h}_n) \rightrightarrows \lambda_1 \underline{\mu}_1$ by (2) and, as λ_2 is light, there is a subsequence of $\{\underline{\mu}_2 \widetilde{h}_n\}$ which is uniformly convergent by 10.1. From this point on, let $\{\underline{\mu}_2 \widetilde{h}_n\}$ designate the uniformly convergent subsequence, and suppose that the limit mapping is $\bar{\mu}_1$. That is,

(5) $\qquad\qquad\qquad \underline{\mu}_2 \widetilde{h}_n \rightrightarrows \bar{\mu}_1.$

Notice that $\underline{\mu}_2 \widetilde{h}_n (H_1) = \underline{\mu}_2 (H_2) = \mathfrak{H}_2$, $n = 1, 2, 3, \cdots$. Hence, $\bar{\mu}_1 : H_1 \rightrightarrows \mathfrak{H}_2$ is monotone by 10.2.

Since $\underline{\mu}_2 \widetilde{h}_n \rightrightarrows \bar{\mu}_1$, it is true that:

(6) $\qquad\qquad\qquad \lambda_2 \underline{\mu}_2 \widetilde{h}_n \rightrightarrows \lambda_2 \bar{\mu}_1.$

But $\lambda_2 \underline{\mu}_2 \widetilde{h}_n = \widetilde{m}_2 \widetilde{h}_n$ by (2), and now $\widetilde{m}_1 = \lambda_2 \bar{\mu}_1$ in virtue of (4). Hence $\lambda_1 \underline{\mu}_1$ and $\lambda_2 \bar{\mu}_1$ are both monotone–light factorizations of $\widetilde{m}_1$. Consequently, by 3.5, there is a unique homeomorphism, $\widetilde{\mathfrak{H}} : \mathfrak{H}_1 \rightrightarrows \mathfrak{H}_2$, such that:

(7) $\qquad$ (i) $\qquad \widetilde{\mathfrak{H}} \underline{\mu}_1 = \bar{\mu}_1.$

$\qquad\qquad$ (ii) $\qquad \lambda_2 \widetilde{\mathfrak{H}} = \lambda_1.$

Hence, in view of (5) and (7, i):

(8) $\qquad\qquad\qquad \underline{\mu}_2 \widetilde{h}_n \rightrightarrows \widetilde{\mathfrak{H}} \underline{\mu}_1.$

An item of prime importance is the possibility of extending the homeomorphism $\widetilde{\mathfrak{H}} : \mathfrak{H}_1 \rightrightarrows \mathfrak{H}_2$, in a suitable manner, to a homeomorphism $\mathfrak{H} : \mathfrak{X}_1 \rightrightarrows \mathfrak{X}_2$.

In this connection, the components of $X_i - H_i$ require examination, $i = 1, 2$. Let $R_{i1}, \cdots, R_{ir}$ be the components of $X_i - H_i$ and suppose $\dot{R}_{ik} = J_{ik}$, $k = 1, \cdots, r$; $i = 1, 2$. Then J_{ik} is a Jordan curve, a component of $\dot{H}_i$, $k = 1, \cdots, r$; $i = 1, 2$. Hence $\widetilde{h}_n$ has meaning on J_{1k} and, indeed, $\widetilde{h}_n(J_{1k})$ must be a Jordan curve which is a component of $\dot{H}_2$, $k = 1, \cdots, r$; $n = 1, 2, 3, \cdots$. This does not imply that $\widetilde{h}_n(J_{1k})$ is the *same* Jordan curve for each value of n. On the other hand, it is easy to see that one may select a subsequence of the sequence $\{\widetilde{h}_n\}$ for which this will be the case. Let this subsequence also be designated by $\{\tilde{h}_n\}$. It may be assumed that $\tilde{h}_n(J_{1k}) = J_{2k}$, $k = 1, \cdots, r$.

Let $\bar{R}_{ik} = C_{ik}$, $k = 1, \cdots, r$; $i = 1, 2$. Then C_{ik} is a 2-cell. Define $\mathfrak{C}_{ik} = \mu_i(C_{ik})$, $\mathfrak{C}_{ik}^\circ = \mu_i(R_{ik}) = \mu_i(C_{ik}^\circ)$, and $\dot{\mathfrak{C}}_{ik} = \mu_i(J_{ik}) = \mu_i(\dot{C}_{ik})$, $k = 1, \cdots, r$; $i = 1, 2$. Then $\mathfrak{C}_{i1}^\circ, \cdots, \mathfrak{C}_{ir}^\circ$ are the components of $\mathfrak{X}_i - \mathfrak{H}_i$, $\mathfrak{C}_{ik} = \mathfrak{C}_{ik}^\circ \cup \dot{\mathfrak{C}}_{ik}$, and $\mathfrak{C}_{ik}^\circ \cap \dot{\mathfrak{C}}_{ik} = 0$, $k = 1, \cdots, r$; $i = 1, 2$. (See (1) and 15.3.)

For each $k = 1, \cdots, r$, let $g_i^k = \mu_i | C_{ik}$, $\dot{g}_i^k = g_i^k | \dot{C}_{ik}$, $\dot{h}_n^k = \tilde{h}_n | \dot{C}_{ik}$, and $\dot{\mathfrak{h}}^k = \tilde{\mathfrak{h}} | \dot{\mathfrak{C}}_{ik}$, $i = 1, 2$. Then $\dot{g}_2^k \dot{h}_n^k \rightrightarrows \dot{\mathfrak{h}}^k \dot{g}_1^k$, and the other hypotheses of lemma 15.5 are fulfilled for $k = 1, \cdots, r$. By r successive applications of the lemma one obtains: an extension of $\dot{\mathfrak{h}}^k$ to a homeomorphism, $\mathfrak{h}^k : \mathfrak{C}_{1k} \approx \mathfrak{C}_{2k}$; a subsequence, $\{\dot{h}_{n_j}^k\}$, of $\{\dot{h}_n^k\}$; and an extension of $\dot{h}_{n_j}^k$ to a homeomorphism, $h_{n_j}^k : C_{1k} \rightrightarrows C_{2k}$, $k = 1, \cdots, r$; $j = 1, 2, 3, \cdots$, such that:

$$(9) \qquad g_2^k h_{n_j}^k \rightrightarrows \mathfrak{h}^k g_1^k, \quad k = 1, \cdots, r.$$

Define

$$\mathfrak{h}(x_1) = \begin{cases} \tilde{\mathfrak{h}}(x_1) & \text{if } x_1 \in \mathfrak{H}_1. \\ \\ \mathfrak{h}^k(x_1) & \text{if } x_1 \in \mathfrak{C}_{1k}, \ k = 1, \cdots, r. \end{cases}$$

And for $j = 1, 2, 3, \cdots$, define:

$$h_{n_j} = \begin{cases} \tilde{h}_{n_j}(x_1) & \text{if } x_1 \in H_1. \\ \\ h_{n_j}^k(x_1) & \text{if } x_1 \in C_{1k}, \ k = 1, \cdots, r. \end{cases}$$

Now $\mathfrak{h} : \mathfrak{X}_1 \approx \mathfrak{X}_2$ is an homeomorphism which is an extension of $\tilde{\mathfrak{h}} : \mathfrak{H}_1 \approx \mathfrak{H}_2$, while $h_{n_j} : X_1 \approx X_2$ is a homeomorphism which is an extension of $\tilde{h}_{n_j} : H_1 \approx H_2$, $j = 1, 2, 3, \cdots$; and in view of (8) and (9),

$$\mu_2 h_{n_j} \rightrightarrows \mathfrak{h} \mu_1.$$

Define $m_1 = \mathfrak{h}\mu_1$, $m_2 = \mu_2$, $\mathfrak{X} = \mathfrak{X}_2$, $\mathfrak{H} = \mathfrak{H}_2$ and $h = \lambda_2$. It follows that:

(i) m_i is weakly admissible with respect to X_i, H_i, $\mathfrak{X}$ and $\mathfrak{H}$, $i = 1, 2$, and $m_1 \sim m_2$.

(ii) h is a homeomorphism from $\mathfrak{H}$ onto $\tilde{\mathfrak{H}}$ such that $\tilde{m}_i = h(m_i | H_i)$, $i = 1, 2$. (Use (2), (3) and (7,ii).)

Remark: It has been shown that R_1^k and R_2^k are corresponding components of $X_1 - H_1$ and $X_2 - H_2$, respectively, $k = 1, \cdots, r$ (see 15.3).

15.7   It is now possible to prove the principal result of this section.

THEOREM:

Given:

1°. *Two Fréchet equivalent mappings,* $f_1 : H_1 \to Y$ *and* $f_2 : H_2 \to Y$, *where* H_1 *and* H_2 *are 2-manifolds with boundary.*

2°. X_i *is a closed 2-manifold obtained from* H_i *by capping the bounding Jordan curves with 2-cells,* $i = 1, 2$.

Conclusion:

3°. *There is a mapping,* m_i, *weakly admissible with respect to* X_i, H_i, $\mathfrak{X}$ *and* $\mathfrak{H}$, $i = 1, 2$, *such that* $m_1 \sim m_2$.

4°. *There is a light mapping,* $l : \mathfrak{H} \to Y$, *such that* $f_i = l(m_i | H_i)$, $i = 1, 2$.

Proof: Note that H_i is withdrawn from X_i. By the reduction theorem (1.8) there is a monotone–light factorization, $\tilde{l}\tilde{m}_i$, of f_i with middle space $\tilde{\mathfrak{H}}$ such that $\tilde{m}_1 \sim \tilde{m}_2$, $i = 1, 2$.

By theorem 15.6 there is a mapping, m_i, weakly admissible with respect to X_i, H_i, $\mathfrak{H}$ and $\mathfrak{X}$, $i = 1, 2$, such that $m_1 \sim m_2$, and a homeomorphism, $h : \mathfrak{H} \approx \tilde{\mathfrak{H}}$, such that $\tilde{m}_i = h(m_i | H_i)$, $i = 1, 2$.

Define $l = \tilde{l}h$, and the theorem follows.

This theorem shows that insofar as necessary conditions for Fréchet equivalence are concerned, one can, in a sense, replace the problem by that of finding necessary conditions for weakly admissible mappings. Consequently, *from this point on the emphasis will be on monotone mappings from closed 2-manifolds.* It should be noted that the converse of this theorem is also true, as will be seen in view of the complete solution of the representation problem. It would be interesting to have a direct proof of this fact.

16.   The case for closed 2-manifolds

It is important to recall that, in view of the remark of 15.2, if $m_1 : X_1 \rightrightarrows \mathfrak{X}$ and $m_2 : X_2 \rightrightarrows \mathfrak{X}$ are monotone mappings from closed 2-manifolds, then m_i is weakly admissible with respect to X_i, X_i, $\mathfrak{X}$ and $\mathfrak{X}$, $i = 1, 2$.

16.1   THEOREM:

Given:

1° *A mapping, m_i, weakly admissible with respect to X_i, H_i, X and $\mathfrak{H}$, $i = 1,\ 2$.*

2° $m_1 \sim m_2$.

Conclusion: *If $\mathfrak{U}$ is any normal region in X, while $U_i = m_i^{-1}(\mathfrak{U})$ and $\underline{m}_i = m_i\,|\,U_i$, $i = 1,\ 2$, then there is a mapping h such that:*

3° $h: U_1 \approx U_2$.

4° *Commutativity holds in the following diagram:*

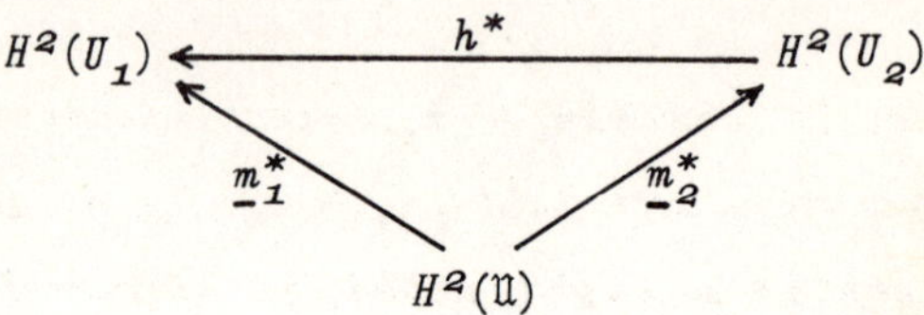

$$H^2(U_1) \xleftarrow{\quad h^* \quad} H^2(U_2)$$
$$\underline{m}_1^* \nwarrow \quad \nearrow \underline{m}_2^*$$
$$H^2(\mathfrak{U})$$

Proof: Suppose M_i is a 2–manifold associated with U_i, $i = 1,\ 2$. In view of 12.4, there is a sequence, $\{\mathfrak{B}_n\}$, of normal regions, $\mathfrak{B}_n$, $n = 1,\ 2,\ 3,\ \cdots$, such that:

$$(1)$$

(i) $\mathfrak{B}_1 \subset \bar{\mathfrak{B}}_1 \subset \mathfrak{B}_2 \subset \bar{\mathfrak{B}}_2 \subset \cdots \subset \mathfrak{U}$.

(ii) $\bigcup \mathfrak{B}_n = \mathfrak{U}$.

(iii) $\dot{\mathfrak{B}}_n$ has the same number of components as $\dot{\mathfrak{U}}$, $n = 1,\ 2,\ 3,\ \cdots$.

(iv) $m_i^{-1}(\mathfrak{B}_n) \supset M_i$, $i = 1,\ 2;\ n = 1,\ 2,\ 3,\ \cdots$.

Let $V_n^i = m_i^{-1}(\mathfrak{B}_n)$, $i = 1,\ 2;\ n = 1,\ 2,\ 3,\ \cdots$. It is easy to see that $V_n^i \approx U_i$, $i = 1,\ 2;\ n = 1,\ 2,\ 3,\ \cdots$. Since $m_1 \sim m_2$ there is a sequence, $\{h_\nu\}$, of homeomorphisms, $h_\nu : X_1 \rightrightarrows X_2$, $\nu = 1,\ 2,\ 3,\ \cdots$, such that $m_2 h_\nu \rightrightarrows m_1$. In view of (1, i), the uniform convergence of $\{m_2 h_\nu\}$ guarantees that there is a sequence of positive integers, $\nu_1 < \nu_2 < \nu < \cdots$, such that for $n = 2,\ 3,\ 4,\ \cdots;\ k \geq \nu_n$:

$$(m_2 h_k)^{-1}(\bar{\mathfrak{B}}_{n-1}) \subset m_1^{-1}(\mathfrak{B}_n) \subset m_1^{-1}(\bar{\mathfrak{B}}_n) \subset (m_2 h_k)^{-1}(\mathfrak{B}_{n+1})$$

To simplify the notation, let $\eta_k = h_{\nu_k}$, $k = 2,\ 3,\ 4,\ \cdots$. Then, for $n = 2,\ 3,\ 4,\ \cdots;\ k \geq n$:

$$m_2^{-1}(\bar{\mathfrak{B}}_{n-1}) \subset \eta_k m_1^{-1}(\mathfrak{B}_n) \subset \eta_k m_1^{-1}(\bar{\mathfrak{B}}_n) \subset m_2^{-1}(\mathfrak{B}_{n+1}),$$

whence

$$(2) \qquad \bar{V}_{n-1}^2 \subset \eta_k(V_n^1) \subset \eta_k(\bar{V}_n^1) \subset V_{n+1}^2 \subset U_2.$$

Consequently, it follows that for each $k = 2,\ 3,\ 4,\ \cdots$, there is a 2-manifold M_k^1 associated with V_k^1 (and hence with U_1) such that:

$$(3) \qquad \bar{V}_{k-1}^1 \subset (M_k^1)^\circ \subset M_k^1 \subset V_k^1,$$

and if $M_k^2 = \eta_k(M_k^1)$, then:

$$\bar{V}_{k-1}^2 \subset (M_k^2)^\circ \subset M_k^2 \subset V_{k+1}^2 \subset U_2.$$

For each $k = 2, 3, 4, \cdots$, it follows that M_k^2 is a 2-manifold associated with V_{k+1}^2, that each component of $U_i - M_k^i$ is an open cylinder, $i = 1, 2$, and that there is a homeomorphism,

$$t_k : U_1 \approx U_2,$$

with

$$t_k \, | \, M_k^1 = \eta_k \, | \, M_k^1.$$

Hence $t_k^* : H^2(U_2) \approx H^2(U_1)$ and it may be assumed that t_k^* is independent of $k = 2, 3, 4, \cdots$, since the groups involved are cyclic infinite or of order 2.

Notice that for fixed $n = 2, 3, 4, \cdots$, if $k \geq n$ and $x_1 \in M_n^1$, then $x_1 \in M_k^1$ whence $t_k(x_1) = \eta_k(x_1)$. Consequently, for fixed $n = 2, 3, 4, \cdots$, and $k \geq n$,

$$(4) \qquad t_k \, | \, M_n^1 = \eta_k \, | \, M_n^1.$$

Since $m_2 \eta_k \rightrightarrows m_1$ on X_1, for fixed $n = 2, 3, 4, \cdots$,

$$[(\underline{m}_2 \eta_k) \, | \, M_n^1] \rightrightarrows [\underline{m}_1 \, | \, M_n^1];$$

that is,

$$(5) \qquad [(\underline{m}_2 t_k) \, | \, M_n^1] \rightrightarrows [\underline{m}_1 \, | \, M_n^1].$$

(Recall that $\underline{m}_i = m_i \, | \, U_i$, $i = 1, 2$.)

For $n = 3, 4, 5, \cdots$, and $k \geq n$,

$$
\begin{aligned}
t_k(U_1 - M_n^1) &= U_2 - t_k(M_n^1) \\
&= U_2 - \eta_k(M_n^1) \quad \text{by (4)} \\
&\subset U_2 - \eta_k(\bar{V}_{n-1}^1) \quad \text{by (3)} \\
&\subset U_2 - V_{n-2}^2 \quad \text{by (2)}.
\end{aligned}
$$

Therefore,

$$(6) \qquad \underline{m}_2 t_k(U_1 - M_n^1) \subset \mathfrak{U} - \mathfrak{V}_{n-2}.$$

On the other hand,

$$
\begin{aligned}
U_1 - M_n^1 &\subset U_1 - \bar{V}_{n-1}^1 \quad \text{by (3)} \\
&\subset U_1 - V_{n-2}^1,
\end{aligned}
$$

hence

(7) $$\underline{m}_1(U_1 - M_n^1) \subset \mathfrak{U} - \mathfrak{B}_{n-2},$$

For notational convenience, let $f_1 = \underline{m}_1$ and $f_k = \underline{m}_2 t_k$, $k = 2, 3, 4, \cdots$. These mappings are compact (see 8.1 and 8.2), hence upon compactifying U_1 and $\mathfrak{U}$ by adding points $\check{u}_1$ and $\check{u}$, one obtains mappings $\check{f}_k : (\check{U}_1, \check{u}_1) \to (\check{\mathfrak{U}}, \check{u})$, $k = 1, 2, 3, \cdots$. The spaces $\check{U}_1$ and $\check{\mathfrak{U}}$ are compacta, and in view of (1, ii), (5), (6) and (7) it follows that:

$$\check{f}_k \rightrightarrows \check{f}_1.$$

Hence by 7.9, if $\mathfrak{u} \in H^2(\check{\mathfrak{U}}, \check{u})$ then there is an integer k_o such that $k > k_o$ implies

$$\check{f}_1^*(\mathfrak{u}) = \check{f}_k^*(\mathfrak{u});$$

that is, regarding $\mathfrak{u}$ as an element of $H^2(\mathfrak{U})$ as in 8.2,

$$f_1^*(\mathfrak{u}) = f_k^*(\mathfrak{u})$$

or

$$m_1^*(\mathfrak{u}) = (\underline{m}_2 t_k)^*(\mathfrak{u}) = t_k^* m_2^*(\mathfrak{u}).$$

But t_k^* is independent of $k = 2, 3, 4, \cdots$, hence if $h = t_2$, then :

$$h : U_1 \approx U_2,$$

and commutativity holds in the following diagram:

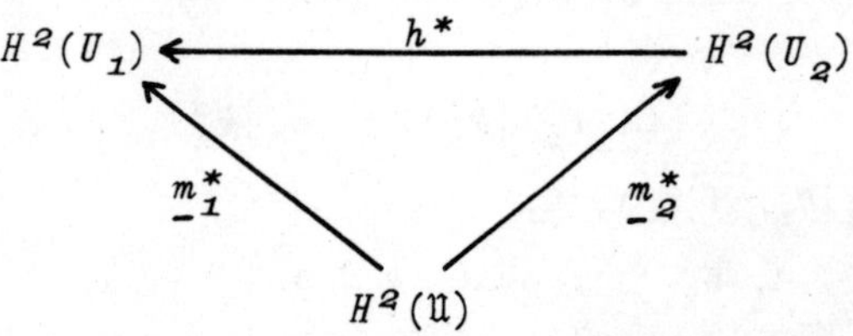

16.2   The theorem of the last paragraph provides a necessary condition for Fréchet equivalence, and the proof has been relatively simple. In the general nature of events, therefore, one rather expects the sufficiency argument to be more difficult. This is indeed the case; consequently the sufficiency theorems will be enhanced if the necessary condition developed in 16.1 can be weakened and still prove sufficient. This is the object of the next two theorems.

16.3   THEOREM:

Given:

1°. *A mapping, m_i, weakly admissible with respect to X_i, H_i, X and $\mathfrak{H}$, $i = 1, 2$.*

2°. *If $\mathfrak{U}$ is any normal region in X, while $U_i = m_i^{-1}(\mathfrak{U})$ and $\underline{m}_i = m_i | U_i$, $i = 1, 2$, then there is a homeomorphism, $h : U_1 \approx U_2$, such that commutativity holds in the following diagram:*

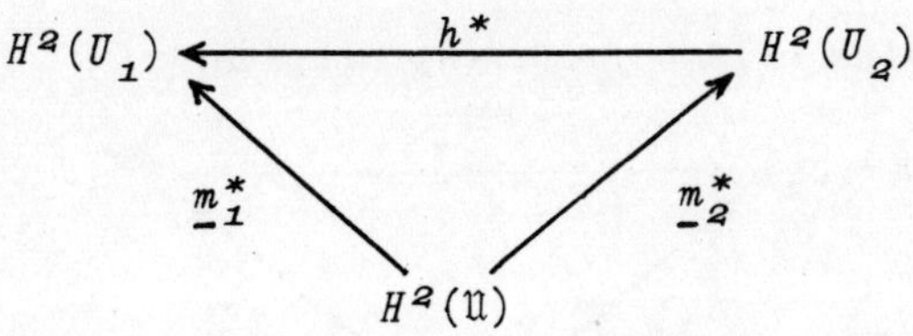

Conclusion: m_1 *and* m_2 *are compatible.*

Proof: Comparability of m_1 and m_2 is obvious.

Relative to consistency, suppose $\mathfrak{U}$ is a simple region in X such that U_1 and U_2 are orientable where $U_i = m_i^{-1}(\mathfrak{U})$. In view of the hypothesis, there is a homeomorphism,

$$h : U_1 \approx U_2,$$

such that commutativity holds in the following diagram:

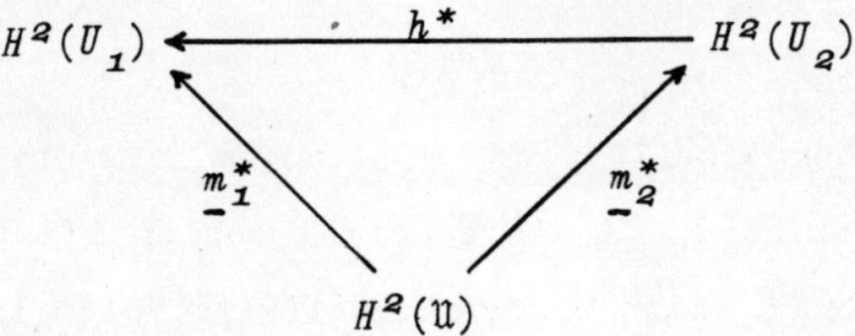

where $\underline{m}_i = m_i | U_i$, $i = 1, 2$.

Let $\{\mathfrak{L}_n\}$ be the set of all true cyclic elements of $\bar{\mathfrak{U}}$, and suppose $r_n : \bar{\mathfrak{U}} \rightrightarrows \mathfrak{L}_n$ is the monotone retraction, $n = 1, 2, 3, \cdots$. If $\underline{r}_n = r_n | \mathfrak{U}$ and $\rho_{in} = (r_n m_i) | U_i$, then $\underline{r}_n^* : H^2(r_n(\mathfrak{U})) \to H^2(\mathfrak{U})$, $\rho_{in}^* : H^2(r_n(\mathfrak{U})) \to H^2(U_i)$, and $\rho_{in}^* = \underline{m}_i^* \underline{r}_n^*$, $i = 1, 2; n = 1, 2, 3, \cdots$. But

$$\underline{m}_1^* = h^* \underline{m}_2^*,$$

whence

$$\underline{m}_1^* \underline{r}_n^* = h^* \underline{m}_2^* \underline{r}_n^*, \quad n = 1, 2, 3, \cdots.$$

Therefore,

$$\rho_{1n}^* = h^* \rho_{2n}^*, \quad n = 1, 2, 3, \cdots,$$

showing that m_1 and m_2 are consistent by 14.6.

16.4   THEOREM:

   Given:

    1°. *A mapping, m_i, weakly admissible with respect to X_i, H_i, $\mathfrak{X}$ and $\mathfrak{H}$, where X_i is orientable, $i = 1, 2$.*

    2°. *If $\mathfrak{U}$ is any normal region in $\mathfrak{X}$, while $U_i = m_i^{-1}(\mathfrak{U})$ and $\underline{m}_i = m_i | U_i$, $i = 1, 2$, then there is a homeomorphism, $h:U_1 \approx U_2$, such that commutativity holds in the following diagram:*

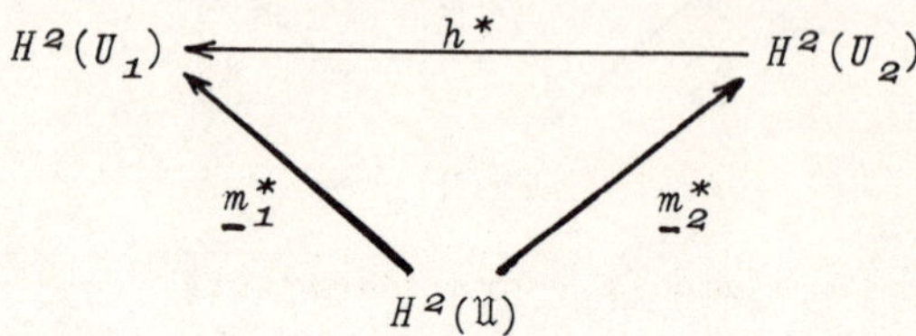

   Conclusion: *m_1 and m_2 are o-comparable and there is an isomorphism, $\eta:H^2(X_2) \approx H^2(X_1)$, such that commutativity holds in the following diagram:*

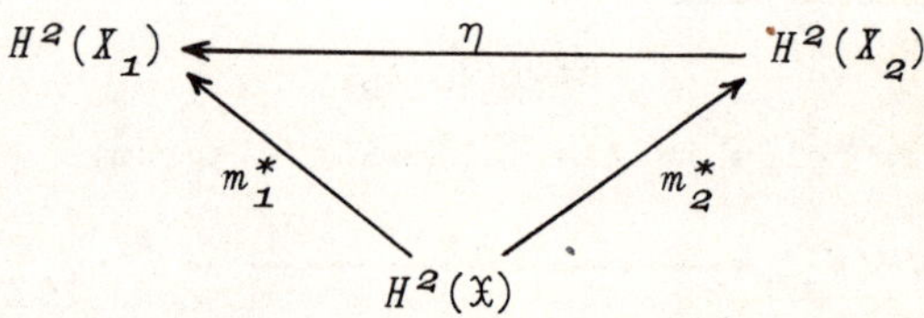

   Proof: O–comparability of m_1 and m_2 is obvious.

   On the other hand, $\mathfrak{X}$ is a normal region of $\mathfrak{X}$. Hence there is a homeomorphism, $h:X_1 \approx X_2$, such that commutativity holds in the following diagram:

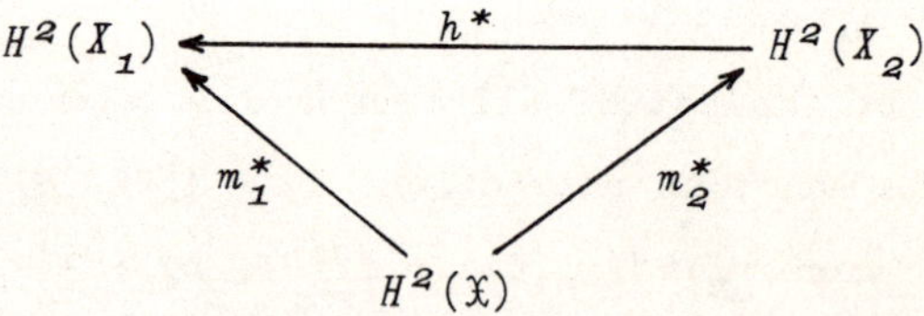

Let $\eta = h^*$ and the theorem is proved.

16.5   THEOREM:

   Given:

    1°. *A mapping m_i, weakly admissible with respect to X_i, H_i, $\mathfrak{X}$ and $\mathfrak{H}$, where X_i is orientable, $i = 1, 2$.*

    2°. *m_1 and m_2 are o–comparable.*

3°. *There is an isomorphism* $\eta : H^2(X_2) \approx H^2(X_1)$ *such that commutativity holds in the following diagram:*

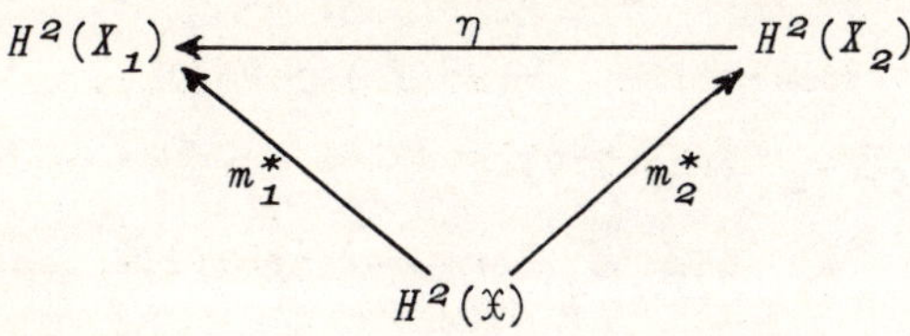

Conclusion: m_1 *and* m_2 *are o-compatible.*

Proof: Suppose $\mathfrak{L}$ is any hyper element of $\mathfrak{X}$, while $\mathfrak{T}$ is the true cyclic element of $\mathfrak{X}$ containing $\mathfrak{L}$ and $r : \mathfrak{X} \rightrightarrows \mathfrak{T}$ is the monotone retraction. Let $\mathfrak{C}$ be a 2-cell on $\mathfrak{L}$ such that $m_i^{-1} r^{-1}(\mathfrak{C}^\circ)$ is an open 2-cell, G_i, $i = 1, 2$. (See 14.9.) Define $\mu_i = (r m_i)|G_i$, and consider the following diagram for $i = 1, 2$:

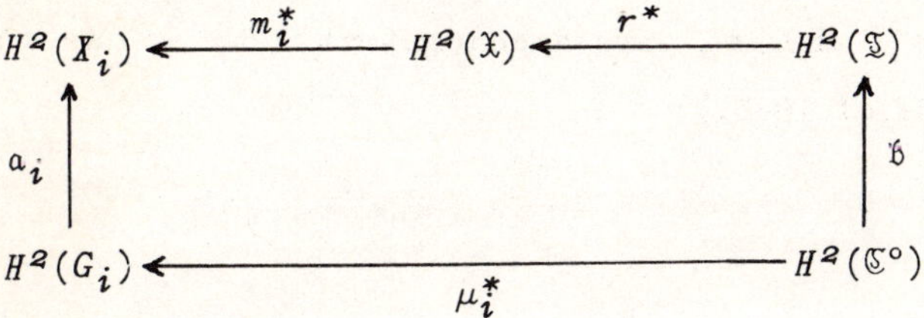

Commutativity holds, and hence:

$$m_i^* r^* \mathfrak{b} = a_i \mu_i^*, \quad i = 1, 2.$$

But

$$m_1^* = \eta m_2^*,$$

whence

$$m_1^* r^* \mathfrak{b} = \eta m_2^* r^* \mathfrak{b}.$$

Therefore,

$$a_1 \mu_1^* = \eta a_2 \mu_2^*.$$

Consequently, m_1 and m_2 are o-consistent by 14.9. Since they are given o-comparable, they are o-compatible.

16.6   In reference to theorems 16.1, 16.4 and 16.5, it is an easy matter to check the following statement which is merely a sample of several possible results.

THEOREM:

Given:

1° *A mapping m_i, weakly admissible with respect to X_i, H_i, $\mathfrak{X}$ and $\mathfrak{H}$, where X_i is orientable, $i = 1, 2$.*

2° *An isomorphism, $\eta : H^2(X_2) \approx H^2(X_1)$.*

3° *$m_1 \overset{\eta}{\sim} m_2$ (see 2.7).*

Conclusion: *m_1 and m_2 are o-compatible, and commutativity holds in the following diagram:*

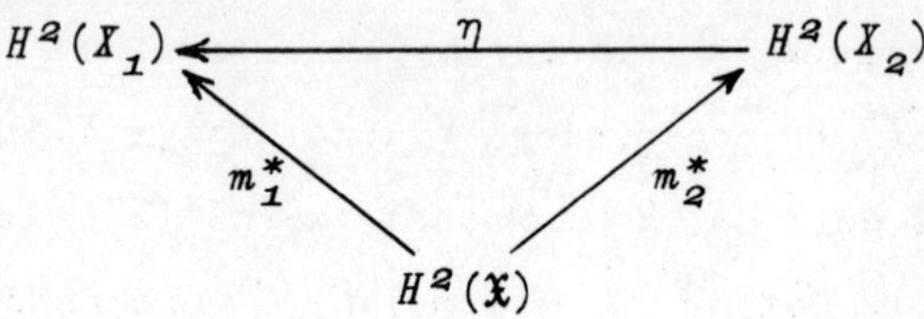

Remark: The second part of the conclusion may be replaced by the statement that m_1 and m_2 are o-consistent with η as a matching isomorphism. (See proof of 16.5.)

CHAPTER V

THE SUFFICIENCY ARGUMENT

It will be observed that the proofs in this direction are by no means as simple as those in chapter IV. This being the case, it is expedient to give an indication of the line of argument before plunging into details.

The first section, as in chapter IV, is essentially concerned with the case for 2-manifolds with boundary. From then on, closed 2-manifolds and 2-manifolds with boundary are handled simultaneously. Relative to the first section, if m_i is *weakly admissible* with respect to X_i, H_i, $\mathfrak{X}$ and $\mathfrak{H}$ (see 15.2), then $m_i|H_i$ is of course monotone, but $m_i|\dot{H}_i$ need not be, $i = 1, 2$. The idea is to modify m_i slightly so as to obtain a new mapping, μ_i, which is *admissible* with respect to X_i, H_i, $\mathfrak{X}$ and a set $\mathfrak{R}$, $i = 1, 2$ (see 15.2). In consequence, $\mu_i|\dot{H}_i$ is not only monotone but a homeomorphism; indeed, $x_i = \mu_i^{-1}\mu_i(x_i)$ if $x_i \in \dot{H}_i$, $i = 1, 2$. This means that the troublesome features which are apt to occur in dealing with $m_i|\dot{H}_i$ disappear on considering $\mu_i|\dot{H}_i$, $i = 1, 2$.

It is important to recall that ultimately one is to be interested in *approximately matching* two mappings (see 1.2). Consequently, it will be equally effective to be able to approximately match two approximations to the two mappings. Hence the fact that m_i must be modified to obtain μ_i is no handicap as long as $\bar{\rho}\{m_i, \mu_i\}$ can be controlled, $i = 1, 2$. These comments provide a natural means of indicating the ideas behind sections 17 and 18.

For the sake of simplicity, suppose $m_1:X_1 \rightrightarrows \mathfrak{X}$ and $m_2:X_2 \rightrightarrows \mathfrak{X}$ are compatible monotone mappings from closed 2-manifolds. Section 19 shows that there are compatible monotone approximations, $\widetilde{m}_1:X_1 \to \mathfrak{X}$ and $\widetilde{m}_2:X_2 \to \mathfrak{X}$ to m_1 and m_2, respectively, such that $\widetilde{m}_1(X_1) = \widetilde{m}_2(X_2)$ and is a *finite mantoid* (see 12.5). Roughly speaking, sections 20 to 22 show that there are monotone approximations, $\mu_1:X_1 \to \mathfrak{X}$ and $\mu_2:X_2 \to \mathfrak{X}$, to $\widetilde{m}_1$ and $\widetilde{m}_2$, respectively, such that $\mu_1(X_1) = \mu_2(X_2) = \widetilde{m}_1(X_1) = \widetilde{m}_2(X_2)$ and there is a factorization,

$\omega_2 \omega_1 h_i$ of μ_i, $i = 1$, 2, such that $h_i : X_i \approx \mathfrak{X}_1$ is a *homeomorphism*, $i = 1$, 2, $\omega_1 : \mathfrak{X}_1 \rightrightarrows \mathfrak{X}_2$ where $\mathfrak{X}_2$ is a *cluster*, (see 12.5), and $\omega_2 : \mathfrak{X}_2 \to \mathfrak{X}$. The sufficiency argument is then completed by using the matching homeomorphism, $(h_2^{-1} h_1) : X_1 \approx X_2$.

17.   The smoothing operation

It has been indicated above that the primary purpose of this section is to replace a mapping, m_i, which is weakly admissible with respect to X_i, H_i, $\mathfrak{X}$ and $\mathfrak{H}$ by a mapping, μ_i, which is admissible with respect to X_i, H_i, $\mathfrak{X}$ and $\mathfrak{R}$ in such a manner that $\bar{\rho}\{m_i, \mu_i\}$ can be controlled, $i = 1$, 2. It is important to notice that this is very easy to do in case all one desires is to obtain a mapping, μ_i, which is admissible with respect to X_i, H_i, $\mathfrak{X}$ and $\mathfrak{R}_i$, $i = 1$, 2. The difficulty is to do the smoothing so that $\mathfrak{R}_1 = \mathfrak{R}_2$. On the other hand, this simultaneous smoothing, so to speak, is indispensable for the applications.

17.1   SMOOTHING THEOREM:

Given, *for* $i = 1$, 2:

1°. *A mapping, m_i, weakly admissible with respect to X_i, H_i, $\mathfrak{X}$ and $\mathfrak{H}$.*

2°. H_i *is withdrawn from* X_i.

3°. $\epsilon > 0$.

Conclusion: *For $i = 1$, 2 there is a homeomorphism, $h_i : X_i \approx X_i$, such that if $\mu_i = m_i h_i$ and $\mu_1(H_1) = \mathfrak{R}$, then:*

4°. μ_i *is admissible with respect to* X_i, H_i, $\mathfrak{X}$ *and* $\mathfrak{R}$.

5°. $\mathfrak{R}^\circ \supset \mathfrak{H}$.

6°. $\bar{\rho}\{m_i, \mu_i\} < \epsilon$.

Proof: Let R_1^k and R_2^k be corresponding components of $X_1 - H_1$ and $X_2 - H_2$, respectively, $k = 1, \cdots, n$ (see 15.3). In other words, $m_1(R_1^k) = \mathfrak{R}^k = m_2(R_2^k)$ where $\mathfrak{R}^k$ is a component of $\mathfrak{X} - \mathfrak{H}$, $k = 1, \cdots, n$, and $\mathfrak{X} - \mathfrak{H} = \bigcup \mathfrak{R}^k$.

For each R_i^k there is a Jordan region, D_i^k, such that:

(1)
    (i) $D_i^k \supset \bar{R}_i^k$, $i = 1$, 2; $k = 1, \cdots, n$.

    (ii) $\bar{D}_i^k \cap \bar{D}_i^j = 0$, $i = 1$, 2; $j \neq k$; j, $k = 1, \cdots, n$.

Notice that $\dot{D}_i^k \subset H_i^\circ$, $i = 1$, 2; $k = 1, \cdots, n$.

To simplify the notation, let R_1 and R_2 be corresponding components of $X_1 - H_1$ and $X_2 - H_2$, respectively, and D_i the Jordan region selected above for R_i, $i = 1, 2$. Then $m_1(R_1) = \Re = m_2(R_2)$ where $\Re$ is a component of $X - \mathfrak{H}$, and $m_1(\dot{R}_1) = \dot{\Re} = m_2(\dot{R}_2)$. It is to be noted that $\dot{R}_i$ is a Jordan curve, J_i, one of the boundary curves of H_i, $i = 1, 2$. The mappings, h_1 and h_2, of the theorem will be determined in terms of certain mappings, g_1 and g_2, now to be defined.

17.2   Two cases arise: I. $\dot{\Re}$ is degenerate; II. $\dot{\Re}$ is non-degenerate.

Case I. $\dot{\Re} = x$, *a single point of* X.

It is easy to see that $\bar{\Re}$ is a 2-sphere true cyclic element, $\mathfrak{S}$, of X, and $m_1(\bar{R}_1) = \mathfrak{S} = m_2(\bar{R}_2)$. Let Ω^* and $\Re^*$ be two Jordan regions in $\Re$ such that:

$$\Re \supset \bar{\Re}^* \supset \Re^* \supset \bar{\Omega}^*,$$

and

$$d(\mathfrak{S} - \Omega^*) < \epsilon.$$

Let $m_i^{-1}(\Omega^*) = Q_i^*$ and $m_i^{-1}(\Re^*) = R_i^*$, $i = 1, 2$. In view of 1° it is clear that Q_i^* and R_i^* are Jordan regions in R_i, $i = 1, 2$, such that:

$$m_2^{-1}m_1(R_1^*) = R_2^*,$$

$$R_i \supset \bar{R}_i^* \supset R_i^* \supset \bar{Q}_i^*, \quad i = 1, 2,$$

and

$$d[m_i(\bar{R}_i^* - Q_i^*)] < \epsilon, \quad i = 1, 2.$$

Since $m_i(J_i) = x$ and m_i is continuous, there is a Jordan region, Q_i, $i = 1, 2$, such that:

$$D_i \supset \bar{Q}_i \supset Q_i \supset \bar{R}_i, \quad i = 1, 2,$$

and

$$d[m_i(\bar{Q}_i - Q_i^*)] < \epsilon, \quad i = 1, 2.$$

17.3   It can be shown from 10.9 that there is a mapping, g_i, $i = 1, 2$, such that:

(i) $g_i : \bar{Q}_i \approx \bar{Q}_i$, $i = 1, 2$.

(ii) g_i is the identity on $\bar{Q}_i^* \cup \dot{Q}_i$, $i = 1, 2$.

(iii) $g_i(\bar{R}_i) = \bar{R}_i^*$, $i = 1, 2$.

It follows that:

$$\text{(iv)} \quad \rho\{m_i(x_i),\ m_i g_i(x_i)\} < \epsilon, \quad x_i \in Q_i, \quad i = 1,\ 2.$$

$$\text{(v)} \quad \text{If} \quad x_i \in \bar{R}_i, \quad \text{then} \quad x_i = (m_i g_i)^{-1} m_i g_i(x_i), \quad i = 1,\ 2.$$

$$\text{(vi)} \quad m_1 g_1(R_1) = m_2 g_2(R_2).$$

17.4   *Case II.* $\dot{\mathfrak{R}}$ *is non-degenerate.*

The task is to determine mappings, g_1 and g_2, similar to those determined in *Case I*, but the non-degeneracy of $\dot{\mathfrak{R}}$ makes the problem somewhat more difficult.

If $\dot{m}_i = m_i | J_i$, then $\dot{m}_i : J_i \rightrightarrows \dot{\mathfrak{R}}$ but need not be monotone, $i = 1,$ 2. Consider $\lambda_i \dot{\mu}_i$, a monotone-light factorization of $\dot{m}_i$ with middle space, Γ_i (see 3.5).

Since J_i is a Jordan curve and $\dot{\mathfrak{R}}$ is non-degenerate, Γ_i is also a Jordan curve by Whyburn [12, p. 165], $i = 1,\ 2.$

Let $\gamma_i \in \Delta_i$ if and only if $\gamma_i \in \Gamma_i$ and $\dot{\mu}_i^{-1}(\gamma_i)$ is non-degenerate, and therefore an arc of J_i, $i = 1,\ 2.$ The set Δ_i is at most denumerable, since otherwise there would be a non-denumerable number of arcs, disjoint in pairs, lying on J_i, $i = 1,\ 2.$

Let $\mathfrak{S} = \lambda_1(\Delta_1) \cup \lambda_2(\Delta_2)$, and note that:

(2)   (i) $\mathfrak{S} \subset \dot{\mathfrak{R}}.$

         (ii) $\mathfrak{S}$ is at most denumerable.

         (iii) $\lambda_i^{-1}(\mathfrak{S}) \subset \Gamma_i$, $i = 1,\ 2.$

         (iv) If $x \in \dot{\mathfrak{R}} - \mathfrak{S}$, then $\dot{m}_i^{-1}(x)$ is totally disconnected, $i = 1,\ 2.$

Let $\Pi_i = \Gamma_i - \lambda_i^{-1}(\mathfrak{S})$, $i = 1,\ 2.$ If $\Gamma_i - \bar{\Pi}_i \neq 0$, then there is a non-degenerate arc, a, in $\Gamma_1 - \bar{\Pi}_1$, $i = 1,\ 2.$ Since λ_1 is light, $\lambda_1(a)$ is a non-degenerate continuum in $\mathfrak{S}$. Hence $\mathfrak{S}$ is non-denumerable in contradiction to (2, ii). This proves that:

(3) $\Gamma_i = \bar{\Pi}_i$, $i = 1,\ 2.$

and it is clear that:

(4) $\pi_i \in \Pi_i$ implies $\dot{\mu}_i^{-1}(\pi_i) \in J_i$, $i = 1,\ 2.$

The mapping λ_i is from a compactum, consequently λ_i is uniformly continuous. Hence, in virtue of (3) and (4), there is a set of intervals, $\beta^0, \cdots, \beta^q (q > 1)$ on Γ_1 such that:

(5)   (i) $\Gamma_1 = \bigcup \beta^j$.

       (ii) $\beta^j \cap \beta^{j+1} = \pi_1^{j+1}$,   a point of Π_1, $j = 0, \cdots, q$; $(j+1)$ taken mod $(q+1)$.

       (iii) $(\beta^j)^\circ \cap (\beta^k)^\circ = 0$, $j \neq k$; j, $k = 0, \cdots, q$.

       (iv) $d[\lambda_1(\beta^j)] < \epsilon$, $j = 0, \cdots, q$.

Now consider the sets $\dot{\mu}_1^{-1}(\beta^j) \equiv \beta_1^j$, $j = 0, \cdots, q$. These are intervals (since $\dot{\mu}_1$ is monotone), and:

(6)   (i) $J_1 = \bigcup \beta_1^j$.

       (ii) $\beta_1^j \cap \beta_1^{j+1} = \dot{\mu}_1^{-1}(\pi^{j+1}) = p_1^{j+1}$,   a single point of J_1, $j = 0, \cdots, q$; $(j+1)$ taken mod $(q+1)$.

       (iii) $(\beta_1^j)^\circ \cap (\beta_1^k)^\circ = 0$, $j \neq k$; j, $k = 0, \cdots, q$.

       (iv) $d[m_1(\beta_1^j)] < \epsilon$, $j = 0, \cdots, q$.

Notice that:

(7) $m_1(p_1^j) \in \dot{\mathfrak{R}} - \mathfrak{C}$, $j = 0, \cdots, q$.

The rôle of Γ_1 and Γ_2, entirely subsidiary, is now complete.

<u>17.5</u>   With regard to the arcs, $\beta_1^\circ, \cdots, \beta_1^q$, select any point, $a_1 \in R_1$, and let $(a_1, p_1^j) \equiv a_1^j$ be an arc with end points a_1 and p_1^j, $j = 0, \cdots, q$, such that:

(8)   (i) $a_1^j \subset R \cup p_1^j$, $j = 0, \cdots, q$.

       (ii) $a_1^j \cap a_1^k = a_1$, $j \neq k$; j, $k = 0, \cdots, q$.

The components of the set $R_1 - \bigcup a_1^j$ are Jordan regions, $S_1^\circ, \cdots, S_1^q$, where the notation is selected, so that $\dot{S}_1^j \supset \beta_1^j$, $j = 0, \cdots, q$.

By 1° it follows that $m_1(a_1^j)$ is an arc in $\bar{\mathfrak{R}}$ having precisely one point, $m_1(p_1^j)$, on $\dot{\mathfrak{R}}$; in fact, $m_1(p_1^j)$ is an end point of $m_1(a_1^j)$, and $m_1(p_1^j) \in \dot{\mathfrak{R}} - \mathfrak{C}$ by (7), $j = 0, \cdots, q$.

If A_1^j is the interior of the arc a_1^j, then in view of the fact that $m_1(p_1^j) \in \dot{\mathfrak{R}} - \mathfrak{C}$ — and hence $\dot{m}_2^{-1} m_1(p_1^j)$ is *totally* disconnected by (2, iv) — a straightforward argument shows that $m_2^{-1} m_1(A_1^j)$ is an arc, $a_2^j \equiv (a_2, p_2^j)$, in $\bar{R}_2$, $j = 0, \cdots, q$. And now it follows that:

(9)   (i) $a_2^j \subset R_2 \cup p_2^j$, $p_2^j \in \dot{R}_2$, $j = 0, \cdots, q$.

       (ii) $m_2(a_2^j) = m_1(a_1^j)$, $j = 0, \cdots, q$.

      (iii) $(m_2^{-1} m_1)\,|\,(\alpha_1^j - p_1^j)$ maps $(\alpha_1^j - p_1^j)$ topologically onto $(\alpha_2^j - p_2^j)$, $j = 0, \cdots, q$.

      (iv) $m_1(p_1^j) = m(p_2^j) \in \dot{\Re} - \mathfrak{E}$, $j = 0, \cdots, q$.

      (v) $\alpha_2^j \cap \alpha_2^k = a_2$,   where $a_2 = m_2^{-1} m_1(a_1)$, $j \neq k$; $j, k = 0, \cdots, q$.

17.6   The components of the set $R_2 - \bigcup \alpha_2^j$ are Jordan regions, $S_2^o, \cdots, S_2^q$, where the notation is selected, so that $m_2^{-1} m_1(S_1^j) = S_2^j$, $j = 0, \cdots, q$. (Recall that $(m_2^{-1} m_1)\,|\,R_1$ is a homeomorphism from R_1 onto R_2 in virtue of 1°.) It follows that if $\beta_2^j = \dot{S}_2^j \cap J_2$, then β_2^j is an arc on J_2, $j = 0, \cdots, q$. and:

(10)   (i) $J_2 = \bigcup \beta_2^j$.

      (ii) $\beta_2^j \cap \beta_2^{j+1} = p_2^{j+1}$, $j = 0, \cdots, q$; $(j + 1)$ taken mod $(q + 1)$.

      (iii) $(\beta_2^j)^\circ \cap (\beta_2^k)^\circ = 0$, $j \neq k$; $j, k = 0, \cdots, q$.

      (iv) $m_1(\beta_1^j) = m_2(\beta_2^j)$, hence $d[m_i(\beta_1^j)] < \epsilon$, $j = 0, \cdots, q$; $i = 1, 2$. (See (6, iv).)

17.7   Recall that D_i is a Jordan region containing $\bar{R}_i$ (see 17.1). Let $\zeta_i^j \equiv (a_i, b_i^j)$, $j = 0, \cdots, q$; $i = 1, 2$, be an arc such that:

(11)   (i) $\zeta_i^j \supset \alpha_i^j$, $j = 0, \cdots, q$; $i = 1, 2$.

      (ii) $b_i^j \in \dot{D}_i$, $j = 0, \cdots, q$; $i = 1, 2$.

      (iii) $\zeta_i^j \subset D_i \cup b_i^j$, $j = 0, \cdots, q$; $i = 1, 2$.

      (iv) $\zeta_i^j \cap \zeta_i^k = a_i$, $j \neq k$; $j, k = 0, \cdots, q$; $i = 1, 2$.

The components of the set $D_i - \bigcup \zeta_i^j$ are Jordan regions, $T_i^o, \cdots, T_i^q$, where the notation is selected so that $T_i^j \supset S_i^j$, $j = 0, \cdots, q$; $i = 1, 2$. It follows that there are Jordan regions Q_i, R_i^* and Q_i^* such that:

(12)   (i) $D_i \supset \bar{Q}_i \supset Q_i \supset \bar{R}_i \supset R_i \supset \bar{R}_i^* \supset R_i^* \supset \bar{Q}_i^* \supset Q_i^* \ni a_i$, $i \doteq 1, 2$.

      (ii) Each of the Jordan curves, $\dot{Q}_i$, $\dot{R}_i^*$ and $\dot{Q}_i^*$, has exactly one point on ζ_i^j, $j = 0, \cdots, q$; $i = 1, 2$.

      (iii) $m_1(R_1^*) = m_2(R_2^*)$ and $m_1(Q_1^*) = m_2(Q_2^*)$.

      (iv) If C_i^j is the Jordan region $T_i^j \cap (Q_i - \bar{Q}_i^*)$, then $d[m_i(\bar{C}_i^j)] < \epsilon$, $j = 0, \cdots, q$; $i = 1, 2$. (See (10, iv).)

       Working with the 2-cells, $\bar{C}_i^j$, it follows that there is a homeomorphism, $g_i : \bar{Q}_i \approx \bar{Q}_i$, $i = 1, 2$, such that:

(13)   (i) g_i is the identity on $Q_i^* \cup \dot{Q}_i$, $i = 1, 2$.

        (ii) $g_i(\bar{R}_i) = \bar{R}_i^*$, $i = 1, 2$.

      (iii) $g_i(\bar{C}_i^j) = \bar{C}_i^j$, $j = 0, \cdots, q$; $i = 1, 2$.

And now it is a simple matter to see, in view of (12, iii and iv), that g_i satisfies all the properties of the homeomorphism, g_i, in 17.3.

17.8   Combining cases I and II, it is seen that if R_1 and R_2 are corresponding components of $X_1 - H_1$ and $X_2 - H_2$, respectively (that is, $m_1(R_1) = m_2(R_2) = \Re$), then mappings g_1 and g_2 can be selected having properties (i) to (vi) of 17.3, regardless of whether $\dot{\Re}$ is degenerate or not.

Thus, for each of the n pairs of corresponding components, R_1^k and R_2^k, of $X_1 - H_1$ and $X_2 - H_2$, respectively, there is an associated collection of Jordan regions, D_1^k, Q_1^k, R_1^k, R_1^{*k}, Q_1^{*k} and D_2^k, Q_2^k, R_2^k, R_2^{*k}, Q_2^{*k}, together with homeomorphisms g_1^k and g_2^k such that:

     (i) $D_i^k \supset \bar{Q}_i^k \supset Q_i^k \supset \bar{R}_i^k \supset R_i^k \supset \bar{R}_i^{*k} \supset R_i^{*k} \supset \bar{Q}_i^{*k}$, $k = 1, \cdots, n$;

$i = 1, 2$.

     (ii) $\bar{D}_i^k \cap \bar{D}_i^j = 0$, $k \neq j$; $k, j = 1, \cdots, n$.

    (iii) g_i^k has the properties (i) to (vi) of 17.3.

For $i = 1, 2$, define:

$$h_i(x_i) = \begin{cases} x_i & \text{if } x_i \notin \bigcup Q_i^k \\ g_i^k(x_i) & \text{if } x_i \in Q_i^k, \; k = 1, \cdots, n. \end{cases}$$

As a direct consequence of the properties of g_i^k, it follows that h_i maps X_i homeomorphically onto itself, $i = 1, 2$, and if $\mu_i = m_i h_i$ while $\mu_1(H_1) = \Re$, then:

     (i) $\Re^\circ \supset \mathfrak{H}$.

     (ii) $\bar{\rho}\{m_i, \mu_i\} < \epsilon$, $i = 1, 2$.

    (iii) μ_i is admissible with respect to X_i, H_i, $\mathfrak{X}$ and $\Re$.

Relative to admissibility, note that if $x_i \in \overline{X_i - H_i}$, then $x_i = (m_i h_i)^{-1} m_i h_i(x_i)$, $i = 1, 2$, while $m_1 h_1(X_i - H_i) = m_1(\bigcup R_1^{*k}) = m_2(\bigcup R_2^{*k}) = m_2 h_2(X_2 - H_2)$. Consequently, $\mu_2(H_2) = m_2 h_2(H_2) = \mathfrak{X} - m_2 h_2(X_2 - H_2) = \mathfrak{X} - m_1 h_1(X_1 - H_1) = m_1 h_1(H_1) = \Re$.

18. The first modification

The purpose of this section is to show that weak admissibility may be replaced by admissibility.

<u>18.1</u>   THEOREM:

Given, *for* $i = 1, 2$:

1°. m_i *is weakly admissible with respect to* X_i, H_i, $\mathfrak{X}$ *and* $\mathfrak{H}$.

2°. m_1 *and* m_2 *are compatible.*

3°. $\epsilon > 0$.

Conclusion: *For* $i = 1, 2$, *there is a set,* $\mathfrak{K} \subset \mathfrak{X}$, *and a mapping,* μ_i, *such that:*

4°. μ_i *is admissible with respect to* X_i, H_i, $\mathfrak{X}$ *and* $\mathfrak{K}$.

5°. $\mathfrak{K}^\circ \supset \mathfrak{H}$.

6°. μ_1 *and* μ_2 *are compatible.*

7°. $\bar{\rho}\{m_i,\ \mu_i\} < \epsilon$.

Proof: If H_1 is not withdrawn from X_1, then H_2 is not withdrawn from X_2, in view of 1°. In this case, let $\mu_i = m_i$, $i = 1, 2$, and the result is obvious.

If H_1 is withdrawn from X_1, then H_2 is withdrawn from X_2. In this case, use the mapping h_i of the smoothing theorem (17.1) to define $\mu_i = m_i h_i$, $i = 1, 2$.

The only portion of the conclusion which is in doubt is the compatibility of μ_1 and μ_2.

Suppose $\mathfrak{U}$ is a normal region in $\mathfrak{X}$, while $U_i = m_i^{-1}(\mathfrak{U})$ and $V_i = \mu_i^{-1}(\mathfrak{U})$, $i = 1, 2$. Since $\mu_i = m_i h_i$, it is clear that $V_i = h_i^{-1}(U_i)$, $i = 1, 2$. But m_1 and m_2 are comparable, hence $U_1 \approx U_2$ and therefore $V_1 \approx V_2$ since h is a homeomorphism.

Suppose that $\mathfrak{U}$ is a simple region in $\mathfrak{X}$ and V_i is orientable; then U_i is clearly orientable, $i = 1, 2$.

Let $\{\mathfrak{C}_n\}$ be the collection of 2-cell true cyclic elements of $\bar{\mathfrak{U}}$, and $r_n : \bar{\mathfrak{U}} \rightrightarrows \mathfrak{C}_n$ the monotone retraction, $n = 1, 2, 3, \cdots$. Since m_1 and m_2 are consistent, there is a matching isomorphism, $\eta : H^2(U_2) \approx H^2(U_1)$, such that if $\underline{m}_i = m_i | U_i$, $i = 1, 2$, then commutativity holds in the following diagram:

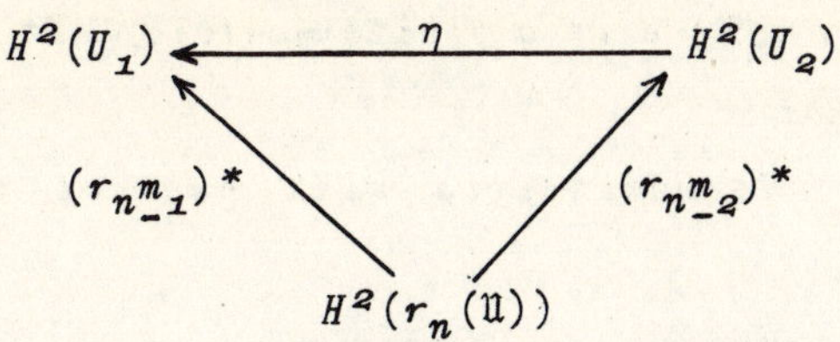

If $\underline{h}_i = h_i | V_i$, then $\underline{h}_i^* : H^2(U_i) \approx H^2(V_i)$, $i = 1, 2$. If $\underline{\mu}_i = \mu_i | V_i$, then $\underline{m}_i = \underline{\mu}_i \underline{h}_i^{-1}$, $i = 1, 2$, and $(r_n \underline{\mu}_1 \underline{h}_1^{-1})^* = \eta (r_n \underline{\mu}_2 \underline{h}_2^{-1})^*$, $n = 1, 2, 3, \cdots$. Consequently, $(\underline{h}_1^{-1})^* (r_n \underline{\mu}_1)^* = \eta (\underline{h}_2^{-1})^* (r_n \underline{\mu}_2)^*$ and $(r_n \underline{\mu}_1)^* = \underline{h}_1^* \eta (\underline{h}_2^*)^{-1} (r_n \underline{\mu}_2)^*$, $n = 1, 2, 3, \cdots$. Since $\underline{h}_1^* \eta (\underline{h}_2^*)^{-1} : H^2(V_2) \approx H^2(V_1)$, the mappings μ_1 and μ_2 are consistent.

18.2   The following theorem may be proved in a similar manner:

THEOREM:

Given, *for* $i = 1, 2$:

1°. m_i *is weakly admissible with respect to* X_i, H_i, $\mathfrak{X}$ *and* $\mathfrak{H}$.

2°. m_1 *and* m_2 *are o–compatible.*

3°. $\epsilon > 0$.

Conclusion: *For* $i = 1, 2$, *there is a set,* $\mathfrak{R} \subset \mathfrak{X}$, *and a mapping,* μ_i, *such that:*

4°. μ_i *is admissible with respect to* X_i, H_i, $\mathfrak{X}$ *and* $\mathfrak{R}$.

5°. $\mathfrak{R}^\circ \supset \mathfrak{H}$.

6°. μ_1 *and* μ_2 *are o–compatible.*

7°. $\overline{\rho}\{m_i, \mu_i\} < \epsilon$.

19.   The second modification

The purpose of this section is to show that the common monotone image of the mappings under consideration (which may be a general mantoid) can be replaced by a finite mantoid (see 12.5).

19.1   THEOREM:

Given:

1°. *A mapping,* m_i, *admissible with respect to* X_i, H_i, $\mathfrak{X}$ *and* $\mathfrak{H}$, $i = 1, 2$.

2°. m_1 *and* m_2 *are compatible.*

3°. $\epsilon > 0$.

Conclusion: *There is a finite mantoid,* $\mathfrak{X}^* \subset \mathfrak{X},$ *and a mapping,*
$\mu_i,$ $i = 1, 2,$ *such that:*

4°. μ_i *is admissible with respect to* $X_i,$ $H_i,$ $\mathfrak{X}^*$ *and*
$\mathfrak{H}^* \equiv \mu_1(H_1),$ $i = 1, 2.$

5°. μ_1 *and* μ_2 *are compatible.*

6°. $\overline{\rho}\{m_i, \mu_i\} < \epsilon,$ $i = 1, 2.$

Proof: Assume that $\mathfrak{X}$ is not a finite mantoid, and let $(\theta, \mathfrak{Y}, \mathfrak{X})$ be a suitable system, where $\mathfrak{P}$ in $\mathfrak{X}$ and Ω in $\mathfrak{Y}$ are the exceptional sets (see 6.1).

If H_i is withdrawn from $X_i,$ $i = 1, 2,$ then $\mathfrak{X} - \mathfrak{H}$ will have r components, $\mathfrak{R}_1,$ $\cdots,$ $\mathfrak{R}_r,$ $r > 0,$ each an open 2-cell with respect to $\mathfrak{X}$. Consequently, $\overline{\mathfrak{R}}_j$ will lie on some true cyclic element of $\mathfrak{X},$ $j = 1,$ $\cdots,$ r. Moreover, $\theta^{-1}|\cup\overline{\mathfrak{R}}_j$ must be a homeomorphism, hence $\theta^{-1}(\mathfrak{R}_j)$ is an open 2-cell in $\mathfrak{Y}$ and consequently lies on some true cyclic element of $\mathfrak{Y},$ $j = 1,$ $\cdots,$ r. Let $\mathfrak{S} \in \mathcal{E}$ if and only if $\mathfrak{S}$ is a true cyclic element of $\mathfrak{Y}$ and $\mathfrak{S} \supset \theta^{-1}(\mathfrak{R}_j)$ for some $j = 1,$ $\cdots,$ r. Notice that $\mathcal{E}$ does not have more than r elements.

If H_i is *not* withdrawn from $X_i,$ $i = 1, 2,$ then $\mathcal{E}$ is empty.

<u>19.2</u>   By 4.3 there is a (possibly finite) sequence, $\{\mathfrak{C}(\mathfrak{p}_k, q_k)\},$ of cyclic chains of $\mathfrak{Y}$ such that, if $\mathfrak{C}_k \equiv \mathfrak{C}(\mathfrak{p}_k, q_k),$ $k = 1, 2, 3, \cdots,$ then:

(1)   (i) $\bigcup_1^n \mathfrak{C}_k \equiv \mathfrak{A}_n$ is an A-set of $\mathfrak{Y},$ $n = 1, 2, 3, \cdots.$

(ii) $\mathfrak{A}_n \cap \mathfrak{C}_{n+1} = q_{n+1},$ $n = 1, 2, 3, \cdots.$

(iii) If δ_n is the maximum of the diameters of the components of $\mathfrak{Y} - \mathfrak{A}_n,$ then $\delta_n \to 0.$ (In the event the sequence is finite, $\delta_n = 0$ for the last value of n.)

(iv) There is an n_0 with the property that $n > n_0$ implies $\mathfrak{A}_n \supset \Omega.$

In view of 6.1 and (1, iii), it is readily seen that there is an $n_1 > n_0$ such that if $n > n_1,$ then $\mathfrak{A}_n$ contains every true cyclic element of $\mathfrak{Y}$ which is not a 2-sphere, and if $\mathfrak{S} \in \mathcal{E},$ then $\mathfrak{A}_n \supset \mathfrak{S}.$

If $n > n_1,$ then $\mathfrak{h} \in \mathfrak{Y} - \mathfrak{A}_n$ implies $\mathfrak{h} \notin \Omega$ and hence $\mathfrak{h} = \theta^{-1}\theta(\mathfrak{h}).$ Thus if $\mathfrak{S}$ is a component of $\mathfrak{Y} - \mathfrak{A}_n,$ then $\theta(\mathfrak{S})$ is a component of $\mathfrak{X} - \theta(\mathfrak{A}_n);$ and if $\mathfrak{S}^*$ is a component of $\mathfrak{X} - \theta(\mathfrak{A}_n),$ then $\theta^{-1}(\mathfrak{S}^*)$ is a component of $\mathfrak{Y} - \mathfrak{A}_n.$

Since θ is continuous on a compactum, it is uniformly continuous. Hence there is a δ such that any set in $\mathfrak{Y}$ of diameter less than δ has an image in $\mathfrak{X}$ of diameter less than $\epsilon/2$. There is an $n_2 > n_1$ such that $n > n_2$ implies $\delta_n < \delta$ (see 1, iii).

If $n > n_2$, then the components of $\mathfrak{X} - \theta(\mathfrak{U}_n)$ form a null sequence and *each has a degenerate frontier.* Consequently, not only is $\mathfrak{U}_n$ an A–set of $\mathfrak{Y}$, but $\theta(\mathfrak{U}_n)$ is an A–set of $\mathfrak{X}$.

Select any $\nu > n_2$, and let

$$(2) \qquad \mathfrak{U} = \mathfrak{U}_\nu = \bigcup_1^\nu \mathfrak{C}(\mathfrak{p}_k, \ q_k),$$
$$(3) \qquad \mathfrak{R} = \mathfrak{Q} \cup \bigcup_1^\nu (\mathfrak{p}_k \cup q_k).$$

If

$$(4) \qquad \mathfrak{U}^* = \theta(\mathfrak{U}),$$

then $\mathfrak{U}^*$ is an A–set of $\mathfrak{X}$ and there is a unique monotone retraction, $\gamma: \mathfrak{X} \rightrightarrows \mathfrak{U}^*$. Hence $\mathfrak{U}^*$ is a mantoid. Moreover, if $\mathfrak{x} \in \mathfrak{U}^*$, then any true cyclic element of $\mathfrak{X}$ in $\gamma^{-1}(\mathfrak{x})$ is a hyper element of $\mathfrak{X}$, and $\mathfrak{x} \in \bigcup \bar{\mathfrak{R}}_j$ implies $\mathfrak{x} = \gamma^{-1}(\mathfrak{x})$. Since each component of $\mathfrak{X} - \mathfrak{U}^*$ has diameter less than $\epsilon/2$,

$$(5) \qquad \rho\{\mathfrak{x}, \ \gamma(\mathfrak{x})\} < \epsilon/2, \ \mathfrak{x} \in \mathfrak{X}.$$

__19.3__ In each cyclic chain, $\mathfrak{C}(\mathfrak{p}_k, \ q_k)$, $k = 1, \cdots, \nu$, select an arc a_k, from $\mathfrak{p}_k$ to q_k, which will be fixed in the remainder of the argument. Let $\mathfrak{x} \in \mathfrak{R}_k$ if and only if $\mathfrak{x}$ separates $\mathfrak{p}_k$ from q_k in $\mathfrak{U}$, and consequently in $\mathfrak{Y}$, $k = 1, \cdots, \nu$. Let $\mathfrak{D}_k = \mathfrak{p}_k \cup \mathfrak{R}_k \cup q_k$. Then $\mathfrak{C}(\mathfrak{p}_k, \ q_k) = \mathfrak{D}_k \cup \bigcup \mathfrak{L}$ where the union is taken over those true cyclic elements, $\mathfrak{L}$, of $\mathfrak{Y}$ such that $\mathfrak{L} \cap \mathfrak{D}_k$ is non–degenerate, $k = 1, \cdots, \nu$ (see Whyburn [10]).

Consequently, each true cyclic element of $\mathfrak{U}$ belongs to exactly one cyclic chain $\mathfrak{C}_k$, $k = 1, \cdots, \nu$. Moreover, if $\mathfrak{L}$ is a true cyclic element in $\mathfrak{C}_k$, then $\mathfrak{L} \cap a_k$ is an arc $a(\mathfrak{L})$ such that $a(\mathfrak{L}) \cap \mathfrak{D}_k$ consists of the end points of $a(\mathfrak{L})$, and $\mathfrak{L} \cap a_j$ is empty or degenerate if $j \neq k$; $j = 1, \cdots, \nu$.

Hence there is a unique arc, $a(\mathfrak{L})$, for each true cyclic element, $\mathfrak{L}$, of $\mathfrak{U}$, and if $\mathfrak{L}_1 \neq \mathfrak{L}_2$, then $a(\mathfrak{L}_1) \cap a(\mathfrak{L}_2)$ is empty or degenerate. Thus there is bi–unique correspondence between the true cyclic elements, $\mathfrak{L}$, and the arcs, $a(\mathfrak{L})$.

$\underline{19.4}$ For $k = 1, \cdots, \nu$, let $\mathfrak{L} \in \mathcal{L}_k$ if and only if:

(6)   (i) $\mathfrak{L}$ is a true cyclic element in $\mathfrak{C}_k$.

(ii) $\mathfrak{L} \cap \mathfrak{N} = 0$ (see (3)). Hence $\mathfrak{L}$ cannot be a nodal true cyclic element of $\mathfrak{C}_k$; that is, $\mathfrak{L} \cap (\mathfrak{C}_k - \mathfrak{L})$ is non-degenerate.

(iii) $\mathfrak{L} \notin \mathfrak{C}$ (see 19.1).

(iv) $\mathfrak{L}$ is a 2-sphere.

(v) $d(\mathfrak{L}) < \delta$.

For $k = 1, \cdots, \nu$, let $\mathfrak{M} \in \mathfrak{m}_k$ if and only if $\mathfrak{M}$ is a true cyclic element of $\mathfrak{C}_k$ and $\mathfrak{M} \notin \mathcal{L}_k$. In view of (6, v), $\mathfrak{m}_k$ consists of a *finite* number of elements (it may be empty). For $k = 1, \cdots, \nu$, let $\mathfrak{B}_k = \bigcup \mathfrak{M}$, $\mathfrak{M} \in \mathfrak{m}_k$ and $\mathfrak{C}_k^* = a_k \cup \mathfrak{B}_k$, then:

(7)   (i) $\mathfrak{C}_k = \mathfrak{C}_k^* \cup \bigcup \mathfrak{L}$, $\mathfrak{L} \in \mathcal{L}_k$.

(ii) $\mathfrak{C}_k^*$ is a cyclic chain relative to itself.

(iii) The set of true cyclic elements of $\mathfrak{C}_k^*$ is $\mathfrak{m}_k$.

(iv) If $\mathfrak{L} \in \mathcal{L}_k$, then $\mathfrak{L} \cap \mathfrak{C}_k^* = a(\mathfrak{L})$, and $d(\mathfrak{L}) < \delta$.

(v) If $\mathfrak{L} \in \mathcal{L}_k$, then $\overline{(\mathfrak{C}_k^* - \mathfrak{L})} \cap \mathfrak{L}$ is the pair of end points of $a(\mathfrak{L})$.

(vi) The components of $[\mathfrak{C}_k^* - (\mathfrak{B}_k \cup \mathfrak{N})]$ are finite in number, and each is an open arc.

Define $\mathcal{L} = \bigcup_1^\nu \mathcal{L}_k$, $\mathfrak{m} = \bigcup_1^\nu \mathfrak{m}_k$, $\mathfrak{B} = \bigcup_1^\nu \mathfrak{B}_k$ and $\mathfrak{Y}^* = \bigcup_1^\nu \mathfrak{C}_k^*$. Then:

(8)   (i) $\mathfrak{A} = \mathfrak{Y}^* \cup \bigcup \mathfrak{L}$, $\mathfrak{L} \in \mathcal{L}$.

(ii) $\mathfrak{Y}^*$ is a generalized cactoid with a cyclic chain decomposition, $\mathfrak{C}_1^*, \cdots, \mathfrak{C}_\nu^*$.

(iii) The set of true cyclic elements of $\mathfrak{Y}^*$ is $\mathfrak{m}$.

(iv) If $\mathfrak{L} \in \mathcal{L}$, then $\mathfrak{L} \cap \mathfrak{Y}^* = a(\mathfrak{L})$, and $d(\mathfrak{L}) < \delta$.

(v) If $\mathfrak{L} \in \mathcal{L}$, then $\overline{(\mathfrak{Y}^* - \mathfrak{L})} \cap \mathfrak{L}$ is the pair of end points of $a(\mathfrak{L})$.

(vi) The components of $[\mathfrak{Y}^* - (\mathfrak{B} \cup \mathfrak{N})]$, if any, are finite in number, and each is an open arc. (In fact, each component of $[\mathfrak{Y}^* - (\mathfrak{B} \cup \mathfrak{N})]$ is a component of $[\mathfrak{C}_k^* - (\mathfrak{B}_k \cup \mathfrak{N})]$ for some k, and for each k any component of $[\mathfrak{C}_k^* - (\mathfrak{B}_k \cup \mathfrak{N})]$ is a component of $[\mathfrak{Y}^* - (\mathfrak{B} \cup \mathfrak{N})]$.)

__19.5__   Suppose $\mathfrak{X}^* = \theta(\mathfrak{Y}^*)$; $\mathfrak{L}^* \in \mathcal{L}^*$ if and only if there is an $\mathfrak{L} \in \mathcal{L}$ such that $\theta(\mathfrak{L}) = \mathfrak{L}^*$; and $\mathfrak{M}^* \in \mathfrak{m}^*$ if and only if there is an $\mathfrak{M} \in \mathfrak{m}$ such that $\theta(\mathfrak{M}) = \mathfrak{M}^*$. If $\mathfrak{L}^* \in \mathcal{L}^*$, then $\mathfrak{L} = \theta^{-1}(\mathfrak{L}^*)$ implies $\mathfrak{L} \in \mathcal{L}$, and $(\theta|\mathfrak{L}):\mathfrak{L} \rightrightarrows \mathfrak{L}^*$ is a homeomorphism by (6, ii). Hence $\theta(a(\mathfrak{L}))$ is an arc, to be designated by $\beta(\mathfrak{L}^*)$, in the 2-sphere $\mathfrak{L}^*$ (see (6, iv)). Moreover:

(9)   (i) $\mathfrak{U}^* = \mathfrak{X}^* \cup \bigcup \mathfrak{L}^*$, $\mathfrak{L}^* \in \mathcal{L}^*$. (See (4).)

(ii) $((\theta|\mathfrak{Y}^*),\ \mathfrak{Y}^*,\ \mathfrak{X}^*)$ is a suitable system.

(iii) The hyper elements of $\mathfrak{X}^*$ are precisely the elements in $\mathfrak{m}^*$, hence they are finite in number.

(iv) If $\mathfrak{L}^* \in \mathcal{L}^*$, then $\mathfrak{L}^* \cap \mathfrak{X}^* = \beta(\mathfrak{L}^*)$, and $d(\mathfrak{L}^*) < \epsilon/2$.

(v) If $\mathfrak{L}^* \in \mathcal{L}^*$, then $\overline{(\mathfrak{X}^* - \mathfrak{L}^*)} \cap \mathfrak{L}^*$ is the pair of end points of $\beta(\mathfrak{L}^*)$.

(vi) The components of $[\mathfrak{X}^* - \{\theta(\mathfrak{B}) \cup \theta(\mathfrak{R})\}]$, if any, are finite in number and each is an open arc. (In fact, each component of $[\mathfrak{X}^* - \{\theta(\mathfrak{B}) \cup \theta(\mathfrak{R})\}]$ is the image, under θ, of a component of $[\mathfrak{Y}^* - (\mathfrak{B} \cup \mathfrak{R})]$, and conversely.)

__19.6__   In view of (9, ii, iii and vi), $\mathfrak{X}^*$ _is a finite mantoid._ If $\mathfrak{L}^* \in \mathcal{L}^*$, it is a 2-sphere and hence there is a monotone retraction,

$$\mu_{\mathfrak{L}^*}:\mathfrak{L}^* \rightrightarrows \beta(\mathfrak{L}^*),$$

such that if $\mathfrak{x}$ is an end point of $\beta(\mathfrak{L}^*)$, then $\mu_{\mathfrak{L}^*}^{-1}(\mathfrak{x}) = \mathfrak{x}$, and if $\mathfrak{x}$ is not an end point of $\beta(\mathfrak{L}^*)$, then $\mu_{\mathfrak{L}^*}^{-1}(\mathfrak{x})$ is a Jordan curve. The retraction is certainly not unique.

For $\mathfrak{x} \in \mathfrak{U}^*$, define:

$$\psi(\mathfrak{x}) = \begin{cases} \mathfrak{x} & \text{if } \mathfrak{x} \notin \bigcup \mathfrak{L}^*,\ \mathfrak{L}^* \in \mathcal{L}^*. \\ \mu_{\mathfrak{L}^*}(\mathfrak{x}) & \text{if } \mathfrak{x} \in \mathfrak{L}^* \in \mathcal{L}^*. \end{cases}$$

Now ψ is a well-defined transformation from $\mathfrak{U}^*$ onto $\mathfrak{X}^*$ by virtue of (9, i, iv and v). It is continuous because the elements in $\mathcal{L}^*$, if infinite in number, form a null sequence, a statement which follows from the fact that the analagous comment holds for $\mathcal{L}$. (See Whyburn [12].) In fact, $\psi:\mathfrak{U}^* \rightrightarrows \mathfrak{X}^*$ is a monotone retraction, for $\mathfrak{x} \in \mathfrak{X}^*$ implies $\psi^{-1}(\mathfrak{x})$ is $\mathfrak{x}$, or a Jordan curve, and

(10) $\rho\{\mathfrak{x},\ \psi(\mathfrak{x})\} < \epsilon/2,\ \mathfrak{x} \in \mathfrak{U}^*.$

Moreover, $x \in \bigcup \overline{\Re}_j$ implies $x = \psi^{-1}(x)$.

Define:

$$(11) \qquad \mu_i = \psi \gamma m_i, \quad i = 1,\ 2.$$

If

$$(12) \qquad \mathfrak{H}^* = \mu_1(H_1),$$

then $\mu_2(H_2) = \psi \gamma m_2(H_2) = \psi \gamma m_1(H_1) = \mu_1(H_1) = \mathfrak{H}^*$. If $x_i \in \overline{X_i - H_i}$, then $\mu_i^{-1}\mu_i(x_i) = m_i^{-1}\gamma^{-1}\psi^{-1}\psi \gamma m_i(x_i)$, $i = 1,\ 2$. But $m_i^{-1}m_i(x_i) = x_i$, hence $m_i(x_i)$ is not a cut point of X, $i = 1,\ 2$. Moreover, $m_i(x_i) \in \bigcup \overline{\Re}_j$, $i = 1,\ 2$. Therefore, $\gamma^{-1}\psi^{-1}\psi \gamma m_i(x_i) = m_i(x_i)$, and $\mu_i^{-1}\mu_i(x_i) = m_i^{-1}m_i(x_i) = x_i$, $i = 1,\ 2$. Hence μ_i is admissible with respect to X_i, H_i, X^* and $\mathfrak{H}^*$.

Finally, $\bar{\rho}\{m_i,\ \mu_i\} = \bar{\rho}\{m_i,\ \psi \gamma m_i\} \leq \bar{\rho}\{m_i,\ \gamma m_i\} + \bar{\rho}\{\gamma m_i,\ \psi \gamma m_i\} < \epsilon$, $i = 1,\ 2$, by (5) and (10).

<u>19.7</u>   Relative to comparability it is a simple matter to see that if V is a normal [simple] region in $\mathfrak{A}^*$ relative to $\mathfrak{A}^*$, then $\mathfrak{B} \equiv \gamma^{-1}(\mathfrak{B})$ is a normal [simple] region in X. (Recall that if $x \in \mathfrak{B}$, then every true cyclic element of X in $\gamma^{-1}(x)$ is a hyper element of X.)

It will *not* be true, in general, that if $\mathfrak{U}$ is a normal [simple] region in X^* with respect to X^*, then $\mathfrak{B} \equiv \psi^{-1}(\mathfrak{U})$ is a normal [simple] region in $\mathfrak{A}^*$ with respect to $\mathfrak{A}^*$. If, however, $\mathfrak{U}$ is a normal [simple] region in X and has the following properties:

$(13) \qquad$ (i) $\mathfrak{U} \subset \beta(\mathfrak{L}^*)$ is false for every $\mathfrak{L}^* \in \mathfrak{L}^*$,

$\qquad$ (ii) If $\mathfrak{L}^* \in \mathfrak{L}^*$ and $\mathfrak{U}$ contains one end point of $\beta(\mathfrak{L}^*)$, then either $\mathfrak{U} \supset \beta(\mathfrak{L}^*)$ or there is a point x in the interior of the arc $\beta(\mathfrak{L}^*)$ such that $x \notin \mathfrak{U}$,

then it follows that $\psi^{-1}(\mathfrak{U})$ is a normal [simple] region in $\mathfrak{A}^*$ with respect to $\mathfrak{A}^*$; hence $\gamma^{-1}\psi^{-1}(\mathfrak{U})$ is a normal [simple] region in X. Consequently, since m_1 and m_2 are comparable, $\mu_1^{-1}(\mathfrak{U}) \approx \mu_2^{-1}(\mathfrak{U})$.

<u>19.8</u>   To consider the first exceptional case, suppose that $\mathfrak{U}$ is a normal region in X^* with respect to X^* contained in $\beta(\mathfrak{L}^*)$ for some $\mathfrak{L}^* \in \mathfrak{L}^*$. Then $\mathfrak{U}$ is an *open arc* in $\beta(\mathfrak{L}^*)$. If $\mathfrak{B} = \psi^{-1}(\mathfrak{U})$, then $\mathfrak{B} = (\mathfrak{L}^*)^\circ$ or $\mathfrak{B}$ is an open cylinder in $(\mathfrak{L}^*)^\circ$, where $(\mathfrak{L}^*)^\circ = \mathfrak{L}^* - \overline{(\mathfrak{A}^* - \mathfrak{L}^*)}$.

If $U_i = \mu_i^{-1}(\mathfrak{U})$, then $\dot{U}_i$ has two components, K_{i_1} and K_{i_2}, and

$\mu_i(K_{i1})$ is one end point, a_1, of the open arc $\mathfrak{U}$, while $\mu_i(K_{i2})$ is the other, a_2, $i = 1, 2$. Moreover, U_i has a single cylinder of approach to K_{ij}, by 12.3, $i, j = 1, 2$.

Suppose $\mathfrak{R}_j = \psi^{-1}(a_j)$, $j = 1, 2$. Let $|\mathfrak{M}_i|$ be the upper semi-continuous decomposition of $\overline{U}_i$ consisting of the sets K_{i1}, K_{i2}, and single points of U_i, $i = 1, 2$. Suppose $\omega_i : \overline{U}_i \rightrightarrows \mathfrak{M}_i$ is the associated mapping, $i = 1, 2$. Then $\mathfrak{M}_i$ is a closed 2-manifold (see 5.3) and $U_1 \approx U_2$ if and only if $\mathfrak{M}_1 \approx \mathfrak{M}_2$. Let $|\mathfrak{R}|$ be the upper semi-continuous decomposition of $\overline{\mathfrak{V}}$ consisting of the sets $\mathfrak{R}_1$, $\mathfrak{R}_2$, and single points of $\mathfrak{V}$, while $\phi : \overline{\mathfrak{V}} \rightrightarrows \mathfrak{R}$ is the associated mapping. Since $\mathfrak{L}^*$ is a 2-sphere, the same is true of $\mathfrak{R}$.

Now consider the monotone mapping, $\xi_i : \mathfrak{M}_i \rightrightarrows \mathfrak{R}$, defined by the formula $\xi_i = \phi \gamma m_i \omega_i^{-1}$, $i = 1, 2$. Suppose $n \in \mathfrak{R}$ and $R(\xi_1^{-1}(n)) \neq 0$. Since U_1 has a single cylinder of approach to K_{11} and K_{12}, the point n cannot be $\phi(\mathfrak{R}_1)$ or $\phi(\mathfrak{R}_2)$. Suppose $\mathfrak{G}$ is the interior of a 2-cell in $[\mathfrak{R} - \{\phi(\mathfrak{R}_1) \cup \phi(\mathfrak{R}_2)\}]$ and $n \in \mathfrak{G}$. Then $\phi^{-1}(\mathfrak{G})$ is the interior of a 2-cell in $[\mathfrak{V} - (\mathfrak{R}_1 \cup \mathfrak{R}_2)]$ and $\gamma^{-1}\phi^{-1}(\mathfrak{G})$ is a simple region in $\mathfrak{X}$. Hence $m_1^{-1}\gamma^{-1}\phi^{-1}(\mathfrak{G}) \approx m_2^{-1}\gamma^{-1}\phi^{-1}(\mathfrak{G})$, since m_1 and m_2 are comparable, and therefore $\xi_1^{-1}(\mathfrak{G}) \approx \xi_2^{-1}(\mathfrak{G})$. It follows that $R(\xi_2^{-1}(n)) \neq 0$. Similarly, if $R(\xi_2^{-1}(n)) \neq 0$, then $R(\xi_1^{-1}(n)) \neq 0$. Hence, by 6.6, there are a finite number of distinct points, $n_1, \cdots, n_c$, $c \geq 0$, for which $R(\xi_i^{-1}(n_k)) \neq 0$, $i = 1, 2$; $k = 1, \cdots, c$, and if $n \neq n_k$, $k = 1, \cdots, c$, then $R(\xi_i^{-1}(n)) = 0$, $i = 1, 2$. Let $\mathfrak{R}_1, \cdots, \mathfrak{R}_c$ be 2-cells on $\mathfrak{R}$ such that:

$$(14) \qquad \text{(i)} \quad \mathfrak{R}_k \cap \mathfrak{R}_j = 0, \ k \neq j; \ k, j = 1, \cdots, c.$$
$$\text{(ii)} \quad \mathfrak{R}_k \cap [\phi(\mathfrak{R}_1) \cup \phi(\mathfrak{R}_2)] = 0, \ k = 1, \cdots, c.$$
$$\text{(iii)} \quad n_k \in \mathfrak{R}_k^{\circ}, \ k = 1, \cdots, c.$$

If $\mathfrak{G}_{ik} = \xi_i^{-1}(\mathfrak{R}_k^{\circ})$, $i = 1, 2$; $k = 1, \cdots, c$, then as above, $\mathfrak{G}_{1k} \approx \mathfrak{G}_{2k}$, $k = 1, \cdots, c$. If $\mathfrak{M}_{ik}$ is a 2-manifold associated with $\mathfrak{G}_{ik}$, $i = 1, 2$; $k = 1, \cdots, c$, then $\mathfrak{M}_{1k} \approx \mathfrak{M}_{2k}$, $k = 1, \cdots, c$. The open set, $\mathfrak{G}_{ik}$, has a single cylinder of approach to the continuum, $\dot{\mathfrak{G}}_{ik}$, hence $\mathfrak{M}_{ik}$ is a 2-manifold with a single boundary curve, Γ_{ik}, $i = 1, 2$; $k = 1, \cdots, c$.

If $\mathfrak{D}_i = \mathfrak{M}_i - \bigcup \mathfrak{M}_{ik}^{\circ}$, then $\mathfrak{D}_i$ is a 2-manifold with boundary curves $\Gamma_{i1}, \cdots, \Gamma_{ic}$; while if $\mathfrak{G}_i$ is a 2-manifold obtained from $\mathfrak{D}_i$ by capping Γ_{ik} with a 2-cell $\mathfrak{G}_{ik}$, $k = 1, \cdots, c$, then it follows that $\mathfrak{G}_i$ is a 2-sphere, $i = 1, 2$. Therefore, $\mathfrak{D}_1 \approx \mathfrak{D}_2$. But $\mathfrak{M}_{1k} \approx \mathfrak{M}_{2k}$, $k = 1, \cdots, c$, and $\mathfrak{M}_i =$

$\mathfrak{D}_i \cup \bigcup \mathfrak{M}_{ik}$, $i = 1, 2$; whence $\mathfrak{M}_1 \approx \mathfrak{M}_2$, and this shows that $U_1 \approx U_2$.

<u>19.9</u> To consider the second exceptional case, suppose $\mathfrak{U}^*$ is a normal [simple] region which contains *one end point* of $\beta(\mathfrak{L}^*)$ for some $\mathfrak{L}^* \in \mathfrak{L}^*$ but $\mathfrak{U}^* \not\supset \beta(\mathfrak{L}^*)$, nor is there a point, x, in the interior of the arc, $\beta(\mathfrak{L}^*)$, such that $x \notin \mathfrak{U}^*$. Then $\mathfrak{U}^* \not\supset \beta(\mathfrak{L}^*)$ but $\mathfrak{U}^* \supset \beta(\mathfrak{L}^*) - a$ where a is the *other end point* of $\beta(\mathfrak{L}^*)$. Hence a is a component of $\dot{\mathfrak{U}}^*$. Therefore, there are but a finite number, $\mathfrak{L}_1^*, \cdots, \mathfrak{L}_c^*$, of distinct elements in $\mathfrak{L}^*$ for which the above situation occurs. Suppose a_k is the end point of $\beta(\mathfrak{L}_k^*)$ not in $\mathfrak{U}^*$, $k = 1, \cdots, c$.

If $U_i^* = \mu_i^{-1}(\mathfrak{U}^*)$, then U_i^* has a single cylinder of approach to each component of $\dot{U}_i^*$ by 6.9, $i = 1, 2$. Let M_i be a 2-manifold associated with U_i^*, $i = 1, 2$. Suppose δ_k is a closed subarc of $\beta(\mathfrak{L}_k^*)$ with a_k for one end point, and such that $\mu_i^{-1}(\delta_k) \cap M_i = 0$, $i = 1, 2$; $k = 1, \cdots, c$. The set $\mathfrak{U} \equiv \mathfrak{U}^* - \bigcup \delta_k$ is a normal [simple] region in $\mathfrak{X}^*$ with respect to $\mathfrak{X}^*$, and $\mathfrak{U}$ has the requirements (i) and (ii) of (13). Hence if $U_i = \mu_i^{-1}(\mathfrak{U}_i)$, then $U_1 \approx U_2$ and $U_i \supset M_i$, $i = 1, 2$. But M_i is clearly a 2-manifold associated with U_i, hence $U_i \approx M_i^o$, $i = 1, 2$. Therefore, $M_1 \approx M_2$ and $U_1^* \approx U_2^*$.

<u>19.10</u> Relative to *consistency*, suppose $\mathfrak{U}^*$ is any *simple* region in $\mathfrak{X}^*$ such that if $U_i^* = \mu_i^{-1}(\mathfrak{U}^*)$, then U_i^* is orientable, $i = 1, 2$. The consistency condition is vacuously fulfilled if $\bar{\mathfrak{U}}^*$ has no true cyclic elements which are 2-cells. Suppose, therefore, that $\bar{\mathfrak{U}}^*$ has at least one true cyclic element which is a 2-cell. In consequence, the first exceptional situation (13, i) cannot arise. If $\mathfrak{U}^*$ does not satisfy property (13, ii), let $\mathfrak{U}$ be a simple region determined as in 19.9; otherwise define $\mathfrak{U} = \mathfrak{U}^*$. Recall that $\mathfrak{U}$ has properties (13, i and ii). Hence if $\mathfrak{W} = \gamma^{-1}\psi^{-1}(\mathfrak{U})$, then $\mathfrak{W}$ is a simple region in $\mathfrak{X}$, while if $U_i = m_i^{-1}(\mathfrak{W})$, then U_i is orientable since $U_i \approx U_i^*$ by 19.9, $i = 1, 2$. Note that each 2-cell true cyclic element of $\bar{\mathfrak{U}}^*$ is also a 2-cell true cyclic element of $\bar{\mathfrak{W}}$.

Suppose $\{\mathfrak{L}_n\}$ is the set of 2-cell true cyclic elements of $\bar{\mathfrak{U}}^*$, while $r_n : \bar{\mathfrak{W}} \rightrightarrows \mathfrak{L}_n$ and $\rho_n : \bar{\mathfrak{U}}^* \rightrightarrows \mathfrak{L}_n$ are the monotone retractions, $n = 1, 2, 3, \cdots$. It is easy to see that $r_n(\mathfrak{W}) = \rho_n(\mathfrak{U}^*)$, $n = 1, 2, 3, \cdots$.

If $\underline{m}_i = m_i | U_i$, $i = 1, 2$, then since m_1 and m_2 are consistent, there is an isomorphism, $\eta : H^2(U_2) \approx H^2(U_1)$, such that commutativity holds in the following diagram for $n = 1, 2, 3, \cdots$:

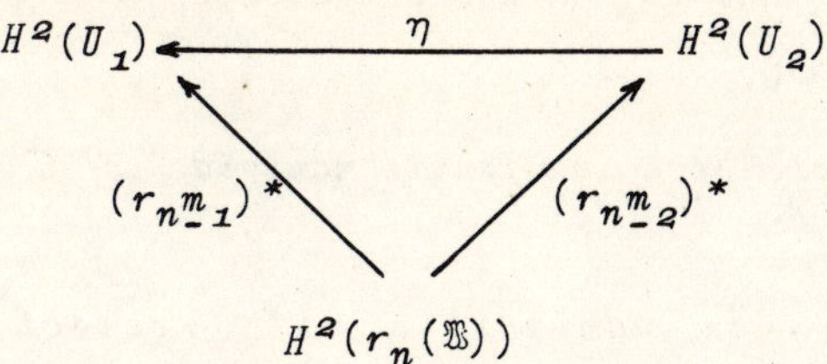

It has been observed that $r_n(\mathfrak{W}) = \rho_n(\mathfrak{U}^*)$, $n = 1, 2, 3, \cdots$, but it is also true that if $\underline{\mu}_i = \mu_i | U_i^*$, $i = 1, 2$, then $(\rho_n\underline{\mu}_i) | U_i = (r_n\underline{m}_i)$. Consequently, one has commutativity in the following diagram of isomorphisms onto, $i = 1, 2$; $n = 1, 2, 3, \cdots$:

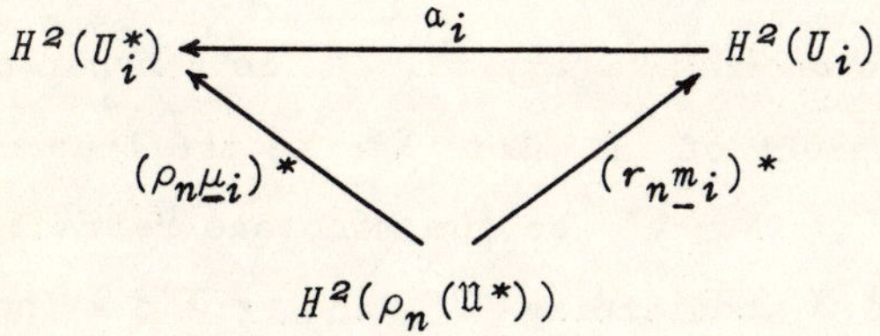

That is, for $i = 1, 2$; $n = 1, 2, 3, \cdots$:

$$(\rho_n\underline{\mu}_i)^* = a_i(r_n\underline{m}_i)^*;$$

hence

$$(r_n\underline{m}_1)^* = \eta(r_n\underline{m}_2)^*$$

implies

$$(\rho_n\underline{\mu}_1)^* = a_1\eta(r_n\underline{m}_2)^*$$
$$= [a_1\eta a_2^{-1}](\rho_n\underline{\mu}_2)^*.$$

If κ is the expression in brackets, then $\kappa : H^2(U_2^*) \approx H^2(U_1^*)$ and for $n = 1, 2, 3, \cdots$,

$$(\rho_n\underline{\mu}_1)^* = \kappa(\rho_n\underline{\mu}_2)^*.$$

Hence μ_1 and μ_2 are consistent.

19.11   In connection with orientable 2—manifolds one has the following result:

THEOREM:

Given:

1° *A mapping, m_i, admissible with respect to X_i, H_i, $\mathfrak{X}$ and*

$\mathfrak{H}$, $i = 1, 2$.

    $2^{\circ}_{\cdot}$ m_1 *and* m_2 *are o—compatible.*

    $3^{\circ}_{\cdot}$ $\eta > 0$.

Conclusion: *There is a finite mantoid,* $\mathfrak{X}^* \subset \mathfrak{X}$, *and a mapping,* μ_i, $i = 1, 2,$ *such that:*

    $4^{\circ}_{\cdot}$ μ_i *is admissible with respect to* X_i, H_i, $\mathfrak{X}$ *and* $\mathfrak{H} = \mu_1(\mathfrak{H}_1)$, $i = 1, 2$.

    $5^{\circ}_{\cdot}$ μ_1 *and* μ_2 *are o—compatible.*

    $6^{\circ}_{\cdot}$ $\bar{\rho}\{m_i, \mu_i\} < \epsilon$, $i = 1, 2$.

Proof: The mappings μ_1 and μ_2 are defined as in 19.6, and the proof of *o*—comparability is the proof of comparability in 19.7 to 19.9.

Relative to *o*—consistency, if $\mathfrak{L}^*$ is a hyper element of $\mathfrak{X}^*$, then it is also a hyper element of $\mathfrak{X}$. Let $\mathfrak{T}^*$ be the true cyclic element of $\mathfrak{X}^*$ containing $\mathfrak{L}^*$, and $\rho : \mathfrak{X}^* \rightrightarrows \mathfrak{T}^*$ be the monotone retraction. Let $\mathfrak{T}$ be the true cyclic element of $\mathfrak{X}$ containing $\mathfrak{L}^*$ and $r : \mathfrak{X} \rightrightarrows \mathfrak{T}$ be the monotone retraction. Suppose $\mathfrak{C}$ is a 2—cell on $\mathfrak{L}^*$ such that $m_i^{-1} r^{-1}(\mathfrak{C}^{\circ})$ is an open 2—cell, G_i, $i = 1, 2$. It follows that $\mu_i^{-1} \rho^{-1}(\mathfrak{C}^{\circ}) = G_i$, and $(r m_i) \,|\, G_i = (\rho \mu_i) \,|\, G_i$, $i = 1, 2$. Since m_1 and m_2 are *o*—consistent, the *o*—consistency of μ_1 and μ_2 is now obvious.

20. The third modification

It is the purpose of this section to show that if the common image of the monotone mappings under consideration is a finite mantoid, then, roughly speaking, it can be replaced by a cluster.

20.1   THEOREM:

    Given:

    $1^{\circ}_{\cdot}$ *A mapping,* m_i, *admissible with respect to* X_i, H_i, $\mathfrak{X}$ *and* $\mathfrak{H}$, $i = 1, 2$.

    $2^{\circ}_{\cdot}$ $\mathfrak{X}$ *is a finite mantoid.*

    $3^{\circ}_{\cdot}$ m_1 *and* m_2 *are compatible.*

    $4^{\circ}_{\cdot}$ $\epsilon > 0$.

Conclusion: *There is a cluster* $\mathfrak{Y}$, *a mapping* $\omega : \mathfrak{Y} \rightrightarrows \mathfrak{X}$, *and a mapping* μ_i, $i = 1, 2,$ *such that:*

$5^{\circ}.$ μ_i *is admissible with respect to* X_i, H_i, $\mathcal{D}$ *and* $\omega^{-1}(\mathfrak{H})$, $i = 1, 2.$

$6^{\circ}.$ μ_1 *and* μ_2 *are compatible.*

$7^{\circ}.$ $\bar{\rho}\{m_i, \omega\mu_i\} < \epsilon$, $i = 1, 2.$

Proof: To avoid trivialities, assume X is non-degenerate but is not a cluster. Since X is a finite mantoid, it has but a finite number of hyper elements, $\mathcal{L}_1, \cdots, \mathcal{L}_l$, and there are a finite number of points, $\mathfrak{p}_1, \cdots, \mathfrak{p}_p$, such that $[X - (\bigcup \mathcal{L}_k \cup \bigcup \mathfrak{p}_k)]$ has a finite number of components, $\beta_1, \cdots, \beta_b$, each an open arc. There is a finite set $\mathfrak{F}$ containing $\bigcup \mathfrak{p}_k$ and such that:

(1)　　　　　　(i) $[X - (\bigcup \mathcal{L}_k \cup \mathfrak{F})]$ has a finite number of components, $a_1, \cdots, a_a$, $a \geqslant 1.$

(ii) a_k is an open arc and $\bar{a}_k$ is an arc in X, $k = 1, \cdots, a.$

(iii) $\dot{a}_j \neq \dot{a}_k$, $j \neq k$; $j, k = 1, \cdots, a.$

(iv) $\mathcal{L}_j \cap \dot{a}_k$ is empty or degenerate, $j = 1, \cdots, l$; $k = 1, \cdots, a.$

(v) $d(a_k) < \epsilon$ and $m_i^{-1}(a_k)$ is an open cylinder, $i = 1, 2$; $k = 1, \cdots, a.$ (See 6.10, and use the fact that m_1 and m_2 are comparable.)

In view of the fact that m_i is admissible, $x_i = m_i^{-1}m_i(x_i)$ for $x_i \in \overline{X_i - H_i}$; hence no point of $m_i(\overline{X_i - H_i})$ is a local cut point of X, and $\bigcup \bar{a}_k \cap m_i(\overline{X_i - H_i}) = 0$, $i = 1, 2.$

Let $\mathfrak{B} = \bigcup \mathcal{L}_k \cup \mathfrak{F} = X - \bigcup a_k$, and $B_i = m_i^{-1}(\mathfrak{B})$, $i = 1, 2.$ Suppose that $|\mathcal{B}_i|$ is the upper semi-continuous decomposition of B_i determined by $m_i | B_i$, that $|\mathcal{D}_i|$ is the completion of $|\mathcal{B}_i|$ over X_i, and that $\phi_i : X_i \rightrightarrows \mathcal{D}_i$ is the associated mapping, $i = 1, 2.$ (See Youngs [12].)

The mapping ϕ_i is monotone, hence $\mathcal{D}_i$ is a mantoid. If, for $i = 1, 2.$

(2)　　　　　　　　　　　$\mathfrak{B}_i = \phi_i(B_i)$

and

(3)　　　　　　　　　　　$\theta_i = m_i \phi_i^{-1},$

then $\theta_i : \mathcal{D}_i \rightrightarrows X$ is monotone and $\theta_i^{-1}|\mathfrak{B}$ is a homeomorphism from $\mathfrak{B}$ onto $\mathfrak{B}_i$, $i = 1, 2.$

It is easy to see that $\theta_i^{-1}(\bar{a}_k)$ is a 2-sphere $\mathfrak{S}_{ik}$, and is a hyper element of $\mathcal{D}_i$, $i = 1, 2$; $k = 1, \cdots, a$; $i = 1, 2$; hence $\mathcal{D}_i$ *is a cluster,*

$i = 1, 2.$

If a_k and b_k are the end points of α_k, consider the points $\theta_i^{-1}(a_k)$ and $\theta_i^{-1}(b_k)$ on the 2–sphere $\mathfrak{S}_{ik}$, $i = 1, 2; k = 1, \cdots, a$. Suppose α_{ik} is the interior of an arc in $\mathfrak{S}_{ij}$ whose end points are $\theta_i^{-1}(a_k)$ and $\theta_i^{-1}(b_k)$, $k = 1, \cdots, a; i = 1, 2.$ Let

$$(4) \qquad\qquad \psi_{ik}: \bar{a}_k \approx \bar{a}_{ik}$$

be a homeomorphism which agrees with θ_i^{-1} on the end points of α_k, $k = 1, \cdots, a; i = 1, 2.$ For $i = 1, 2,$ define:

$$(5) \qquad\qquad \psi_i(x) = \begin{cases} \theta_i^{-1}(x) & \text{if } x \in \mathfrak{B}. \\[2ex] \psi_{ik}(x) & \text{if } x \in a_k, \ k = 1, \cdots, a. \end{cases}$$

Now ψ_i is a homeomorphism from $\mathfrak{X}$ *into* $\mathfrak{Y}_i$ and so imbeds $\mathfrak{X}$ in $\mathfrak{Y}_i$, $i = 1, 2.$ Let

$$(6) \qquad\qquad \mathfrak{X}_i = \psi_i(\mathfrak{X}), \ i = 1, 2.$$

The mapping,

$$(7) \qquad\qquad \gamma_0 \equiv \psi_1 \psi_2^{-1}$$

is a homeomorphism from $\mathfrak{X}_2$ onto $\mathfrak{X}_1$. If

$$(8) \qquad\qquad m_{io} = \psi_i m_i,$$

then $m_{io}: I_i \rightrightarrows \mathfrak{X}_i$, $i = 1, 2.$ Now let $\mu_{20} = \gamma_0 m_{20}$ and notice that $\mu_{20} = \psi_1 \psi_2^{-1} \psi_2 m_2 = \psi_1 m_2.$

Since ψ_1 is a homeomorphism, it is easy to see that m_{10} is admissible with respect to I_1, H_1, $\mathfrak{X}_1$ and $\psi_1(\mathfrak{H})$, while μ_{20} is admissible with respect to I_2, H_2, $\mathfrak{X}_1$ and $\psi_1(\mathfrak{H})$. For the same reason it follows readily that m_{10} and μ_{20} are compatible.

<u>20.2</u>   So as to afford a motivation for the rest of the argument, it is stated that the cluster $\mathfrak{Y}$ of the theorem is to be $\mathfrak{Y}_1$, that $\omega = \theta_1$, that $\mu_1 = \phi_1$, and that the major effort from this point on is to show that the homeomorphism $\gamma_0: \mathfrak{X}_2 \approx \mathfrak{X}_1$ can be extended to a homeomorphism $\gamma: \mathfrak{Y}_2 \approx \mathfrak{Y}_1$ (necessarily mapping $\mathfrak{S}_{2k}$ onto $\mathfrak{S}_{1k}$, $k = 1, \cdots, a$) which has the property that if $\mu_2 = \gamma \phi_2$, then μ_1 and μ_2 are compatible. (It is interesting to note that ω is monotone.)

To this end, suppose $\gamma: \mathfrak{Y}_2 \approx \mathfrak{Y}_1$ is *any* homeomorphic extension of

$\gamma_0 : \mathfrak{X}_2 \approx \mathfrak{X}_1$. Then $\gamma(\mathfrak{S}_{2k}) = \mathfrak{S}_{1k}$, $k = 1, \cdots, a$. If $x_2 \in B_2$, then $\omega\gamma\phi_2(x_2) = m_1\phi_1^{-1}\gamma_0\phi_2(x_2)$ since $\phi_2(x_2) \in \mathfrak{B}_2 \subset \mathfrak{X}_2$. Hence $\omega\gamma\phi_2(x_2) = m_1\phi_1^{-1}\psi_1\psi_2^{-1}\phi_2(x_2) = m_1\phi_1^{-1}\phi_1 m_1^{-1}m_2\phi_2^{-1}\phi_2(x_2) = m_2(x_2)$. On the other hand, $m_2(m_2^{-1}(\bar{a}_k)) = \bar{a}_k$ and $\omega\gamma\phi_2(m_2^{-1}(\bar{a}_k)) = m_1\phi_1^{-1}\gamma\phi_2(m_2^{-1}(\bar{a}_k)) = m_1\phi_1^{-1}\gamma(\mathfrak{S}_{2k}) = m_1\phi_1^{-1}(\mathfrak{S}_{1k}) = \bar{a}_k$, $k = 1, \cdots, a$. Consequently, $\bar{\rho}\{m_2, \omega\gamma\phi_2\} \leqq \max d(\alpha_k) < \epsilon$, whereas $\bar{\rho}\{m_1, \omega\phi_1\} = 0$. This takes care of item 7°.

It is equally simple to show that ϕ_1 is admissible with respect to X_1, H_1, $\mathfrak{Y}$ and $\omega^{-1}(\mathfrak{H})$, while $\gamma\phi_2$ is admissible with respect to X_2, H_2, $\mathfrak{Y}$ and $\omega^{-1}(\mathfrak{H})$.

Hence, to complete the proof, it is only necessary to determine γ so that ϕ_1 and $\gamma\phi_2$ are compatible.

20.3   Suppose first that $\mathfrak{X}$ *has no hyper elements* — hence the same is true of $\mathfrak{X}_1$ and $\mathfrak{X}_2$.

Consider the space,

$$(9) \qquad \mathfrak{X}_i^* = \mathfrak{X}_i \cup \mathfrak{S}_{i1}, \quad i = 1, 2.$$

For $i = 1, 2$, define:

$$(10) \qquad m_{i1}(x_i) = \begin{cases} m_{io}(x_i) & \text{if } x_i \notin \phi_i^{-1}(\mathfrak{S}_{i1}). \\[2mm] \phi_i(x_i) & \text{if } x_i \in \phi_i^{-1}(\mathfrak{S}_{i1}). \end{cases}$$

$$(11) \qquad \rho_i = m_{io}m_{i1}^{-1}$$

Notice that $m_{i1} : \mathit{X}_i \rightrightarrows \mathfrak{X}_i^*$ and $\rho_i : \mathfrak{X}_i^* \rightrightarrows \mathfrak{X}_i$ are monotone, $i = 1, 2$. The mapping ρ_i need not be a retraction, but is the identity on $\mathfrak{B}_i$, $i = 1, 2$.

The mapping γ_0 of (7) carries the subspace $\mathfrak{X}_2$ of $\mathfrak{X}_2^*$ homeomorphically onto the subspace $\mathfrak{X}_1$ of $\mathfrak{X}_1^*$. In particular, it maps the arc $\bar{a}_{21}$ in $\mathfrak{S}_{21}$ homeomorphically onto the arc $\bar{a}_{11}$ of $\mathfrak{S}_{11}$. Extend γ_0 in any manner merely so as to obtain a homeomorphism, $\gamma_1 : \mathfrak{X}_2^* \approx \mathfrak{X}_1^*$.

Define:

$$(12) \qquad \mu_{21} = \gamma_1 m_{21}.$$

It is a simple matter to see that m_{11} is admissible with respect to X_1, H_1, $\mathfrak{X}_1^*$ and $(\psi_1(\mathfrak{H}) \cup \mathfrak{S}_{11})$ while μ_{21} is admissible with respect to X_2, H_2, $\mathfrak{X}_1^*$ and $(\psi_1(\mathfrak{H}) \cup \mathfrak{S}_{11})$.

Assertion: m_{11} *and* μ_{21} *are compatible.*

If $\mathfrak{U}_1$ is a normal region in $\mathfrak{X}_1^*$, while $U_1 = m_{11}^{-1}(\mathfrak{U}_1)$ and $U_2 = \mu_{21}^{-1}(\mathfrak{U}_1)$, then two cases can arise:

$$\text{Case } A. \quad \mathfrak{U}_1 \supset \mathfrak{S}_{11}.$$
$$\text{Case } B. \quad \mathfrak{U}_1 \not\supset \mathfrak{S}_{11}.$$

20.4   *Case A.* It is quite clear that $\rho_1(\mathfrak{U}_1)$ is a normal region in $\mathfrak{X}_1$; moreover, $m_{10}^{-1}[\rho_1(\mathfrak{U}_1)] = m_{11}^{-1}(\mathfrak{U}_1) = U_1$ while $\mu_{20}^{-1}[\rho_1(\mathfrak{U}_1)] = \mu_{21}^{-1}(\mathfrak{U}_1) = U_2$. Hence $U_1 \approx U_2$ since m_{10} and μ_{20} are comparable.

If $\mathfrak{U}_1$ is a simple region in $\mathfrak{X}_1^*$, while U_1 and U_2 are orientable, then $\bar{\mathfrak{U}}_1$ has a single true cyclic element, namely the 2-sphere $\mathfrak{S}_{11}$. Since there are no 2-cell true cyclic elements, the consistency requirement on $\mathfrak{U}_1$ is vacuously fulfilled.

20.5   *Case B.* If $\mathfrak{U}_1 \cap \mathfrak{S}_{11} = 0$, then $\mathfrak{U}_1$ is a normal region of $\mathfrak{X}_1$ and $U_1 = m_{10}^{-1}(\mathfrak{U}_1) \approx \mu_{20}^{-1}(\mathfrak{U}_1) = U_2$. If $\mathfrak{U}_1$ is a simple region of $\mathfrak{X}^*$, while U_1 and U_2 are orientable, then $\bar{\mathfrak{U}}_1$ has no true cyclic elements, and the consistency requirement is vacuously fulfilled.

Suppose, therefore, that $\mathfrak{U}_1 \not\supset \mathfrak{S}_{11}$ but $\mathfrak{U}_1 \cap \mathfrak{S}_{11} \neq 0$. Then $\bar{\mathfrak{U}}_1 \not\supset \mathfrak{S}_{11}$, for if $\bar{\mathfrak{U}}_1 \supset \mathfrak{S}_{11}$ then some point component of $\dot{\mathfrak{U}}_1$ is not an end point of $\bar{\mathfrak{U}}$. (See 2.12.)

Let a_{11} and a_{12} be the end points of α_{11}. Four situations can now arise:

$$\text{(i)} \quad \mathfrak{U}_1 \cap (a_{11} \cup a_{12}) = 0.$$
$$\text{(ii)} \quad \mathfrak{U}_1 \cap (a_{11} \cup a_{12}) = a_{11}.$$
$$\text{(iii)} \quad \mathfrak{U}_1 \cap (a_{11} \cup a_{12}) = a_{12}.$$
$$\text{(iv)} \quad \mathfrak{U}_1 \cap (a_{11} \cup a_{12}) = a_{11} \cup a_{12}.$$

20.6   (i) In this case, $\mathfrak{U}_1$ is the interior of a 2-cell in $\mathfrak{S}_{11}^{\circ} = \mathfrak{S}_{11} - (a_{11} \cup a_{12})$ while $U_1 \approx U_2$ because both are open 2-cells since m_{11}^{-1} and μ_{21}^{-1} are homeomorphisms on $\mathfrak{S}_{11}^{\circ}$. Since $\bar{\mathfrak{U}}_1$ has a single true cyclic element, $\bar{\mathfrak{U}}_1$, m_{11} and μ_{21} are obviously consistent on $\mathfrak{U}_1$ with respect to $\bar{\mathfrak{U}}_1$.

(ii) There is an open subarc, β_{11} of a_{11}, with a_{11} for one end point such that if $\mathfrak{U}_1^* = (\mathfrak{U}_1 - \mathfrak{S}_{11}) \cup \beta_{11} \cup a_{11}$, then $\mathfrak{U}_1^*$ is a normal region of $\mathfrak{X}_1$, and

$$U_1^* = m_{10}^{-1}(\mathfrak{U}_1^*) \subset U_1,$$

$$U_2^* \equiv \mu_{20}^{-1}(\mathfrak{U}^*) \subset U_2.$$

Since m_{10} and μ_{20} are comparable, $U_1^* \approx U_2^*$. Clearly, $\bar{\mathfrak{U}}_1 \cap \mathfrak{S}_{11}$ is a 2–cell true cyclic element, $\mathfrak{S}$, of $\bar{\mathfrak{U}}_1$, and $a_{11} \in \mathfrak{S}^\circ$. Since $m_{11}^{-1}(\mathfrak{S}^\circ - a_{11})$, $\mu_{21}^{-1}(\mathfrak{S}^\circ - a_{11})$, $m_{10}^{-1}(\beta_{11})$ and $\mu_{20}^{-1}(\beta_{11})$ are open cylinders, it follows that $U_i \approx U_i^*$, $i = 1, 2$. Therefore, $U_1 \approx U_2$.

The consistency question is dealt with as in (i).

(iii)   This case is handled in the manner of (ii).

<u>20.7</u>   (iv) If $\mathfrak{U}_1 \cap (a_{11} \cup a_{12}) = a_{11} \cup a_{12}$, then $\bar{\mathfrak{U}}_1$ may have a single true cyclic element, $\mathfrak{S}$, or $\bar{\mathfrak{U}}_1$ may have two 2–cell true cyclic elements, $\mathfrak{S}_1$ and $\mathfrak{S}_2$.

In the first instance, $\mathfrak{S} = \mathfrak{S}_{11} \cap \bar{\mathfrak{U}}_1$ and $(a_{11} \cup a_{12}) \subset \mathfrak{S}^\circ$. Therefore $U_1 = m_{10}^{-1}[\rho_1(\mathfrak{U}_1)] - m_{11}^{-1}(\mathfrak{S}_{11} - \mathfrak{U}_1)$ and $U_2 = \mu_{20}^{-1}[\rho_1(\mathfrak{U}_1)] - \mu_{21}^{-1}(\mathfrak{S}_{11} - \mathfrak{U}_1)$. But $m_{10}^{-1}[\rho_1(\mathfrak{U}_1)] \approx \mu_{20}^{-1}[\rho_1(\mathfrak{U}_1)]$ as in case A, while the sets subtracted are 2–cells. Hence $U_1 \approx U_2$. As for consistency (in case $\mathfrak{U}_1$ is a simple region while U_1 and U_2 are orientable), this is obvious since $\bar{\mathfrak{U}}_1$ has a single true cyclic element.

In the second instance, $\mathfrak{S}_1$ and $\mathfrak{S}_2$ are the components of $\mathfrak{S}_{11} \cap \bar{\mathfrak{U}}_1$ and it may be assumed that $a_{1i} \in \mathfrak{S}_i^\circ$, $i = 1, 2$. The proof that $U_1 \approx U_2$ is an obvious modification of the argument used in (ii).

The consistency proof, in the event $\mathfrak{U}_1$ is a simple region while U_1 and U_2 are orientable, is more involved.

If $\mathfrak{S}_k$ is a 2–cell in $\mathfrak{S}_k^\circ - a_{1k}$, then $m_{11}^{-1}(\mathfrak{S}_k)$ and $\mu_{21}^{-1}(\mathfrak{S}_k)$ are 2–cells, $k = 1, 2$. Let $G_{1k} = m_{11}^{-1}(\mathfrak{S}_k^\circ)$, $G_{2k} = \mu_{21}^{-1}(\mathfrak{S}_k^\circ)$, $k = 1, 2$.

Now consider the following diagram:

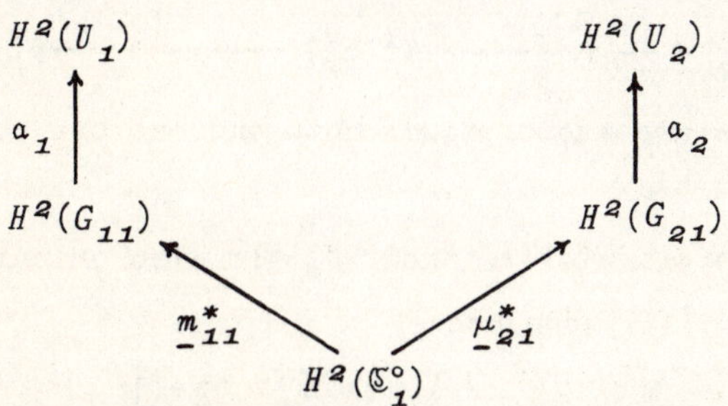

Since all the homomorphisms are isomorphisms onto, there is certainly an isomorphism, $\eta : H^2(U_2) \approx H^2(U_1)$, such that:

(13)
$$a_1 m^*_{\underline{1}1} = \eta a_2 \mu^*_{\underline{2}1}.$$

If $\mathfrak{U}^*_1 = \mathfrak{U}_1 \cup \mathfrak{S}_{11}$, then the formulas $m^{-1}_{11}(\mathfrak{U}^*_1) = m^{-1}_{10}\rho_1(\mathfrak{U}^*_1)$ and $\mu^{-1}_{21}(\mathfrak{U}^*_1) = \mu^{-1}_{20}\rho_1(\mathfrak{U}^*_1)$ hold as in 20.4, while $\rho_1(\mathfrak{U}^*_1)$ is a *normal* (not simple) region of $\mathfrak{X}_1$. Moreover, the comparability of m_{10} and μ_{20} implies that:

$$U^*_1 \equiv m^{-1}_{11}(\mathfrak{U}^*_1) \approx \mu^{-1}_{21}(\mathfrak{U}^*_1) \equiv U^*_2.$$

There is an open cylinder, $\mathfrak{N}$, in $\mathfrak{S}^{\circ}_{11}$ containing $\mathfrak{C}_1$ and $\mathfrak{C}_2$, and such that $\bar{\mathfrak{N}} \cap \mathfrak{C}_k$ is a cylinder, $k = 1, 2$. Note that $\mathfrak{N} \cap \mathfrak{U}_1$ has two components, one containing $\mathfrak{C}_1$ and the other $\mathfrak{C}_2$. Let $N_1 = m^{-1}_{11}(\mathfrak{N})$ and $N_2 = \mu^{-1}_{21}(\mathfrak{N})$, and consider the following diagram for $i = 1, 2$:

$$
\begin{array}{ccc}
H^2(G_{i1}) & \xrightarrow{\;\;c_i\;\;} & H^2(N_i) \\
{\scriptstyle a_i}\big\downarrow & & \big\uparrow{\scriptstyle \mathfrak{b}_i} \\
H^2(U_i) & \xleftarrow{\;\;\mathfrak{b}_i\;\;} & H^2(G_{i2})
\end{array}
$$

All the homomorphisms are isomorphisms onto, and in view of 9.7, one has commutativity [negative commutativity] in the above diagram if U^*_i is orientable [non-orientable], $i = 1, 2$.

Now consider:

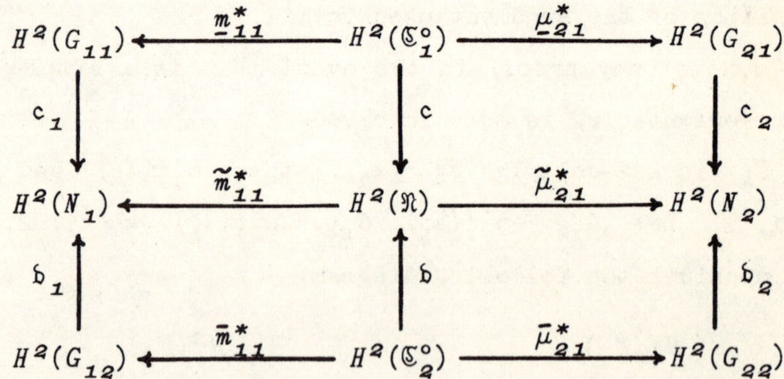

$$
\begin{array}{ccccc}
H^2(G_{11}) & \xleftarrow{\;m^*_{11}\;} & H^2(\mathfrak{S}^{\circ}_1) & \xrightarrow{\;\mu^*_{21}\;} & H^2(G_{21}) \\
{\scriptstyle c_1}\big\downarrow & & {\scriptstyle c}\big\downarrow & & \big\downarrow{\scriptstyle c_2} \\
H^2(N_1) & \xleftarrow{\;\tilde{m}^*_{11}\;} & H^2(\mathfrak{N}) & \xrightarrow{\;\tilde{\mu}^*_{21}\;} & H^2(N_2) \\
{\scriptstyle \mathfrak{b}_1}\big\uparrow & & {\scriptstyle \mathfrak{b}}\big\uparrow & & \big\uparrow{\scriptstyle \mathfrak{b}_2} \\
H^2(G_{12}) & \xleftarrow{\;\bar{m}^*_{11}\;} & H^2(\mathfrak{S}^{\circ}_2) & \xrightarrow{\;\bar{\mu}^*_{21}\;} & H^2(G_{22})
\end{array}
$$

All the homomorphisms are isomorphisms onto and one has commutativity in each box.

If U^*_1 is *orientable*, then U^*_2 is also orientable, and using the last two diagrams and (13), one has:

$$
\begin{aligned}
\mathfrak{b}_1 \bar{m}^*_{11} &= a_1[(c_1)^{-1}\mathfrak{b}_1\bar{m}^*_{11}] \\
&= [a_1 m^*_{\underline{1}1}]c^{-1}\mathfrak{b} \\
&= \eta a_2[\mu^*_{\underline{2}1}c^{-1}\mathfrak{b}]
\end{aligned}
$$

$$= \eta [a_2 c_2^{-1} b_2] \bar{\mu}_{21}^*$$

$$= \eta b_2 \bar{\mu}_{21}^*$$

If U_1^* is *non-orientable*, then U_2^* is also non-orientable, and the only changes which must be made in the array of equalities above is to place a negative sign immediately following the first four equality signs. Hence, again:

$$b_1 \bar{m}_{11}^* = \eta b_2 \bar{\mu}_{21}^*,$$

and so, m_{11} and μ_{21} are consistent on $\mathfrak{U}_1$ with respect to $\mathfrak{S}_1$ and $\mathfrak{S}_2$. (See 14.3.)

This completes the proof of the compatibility of m_{11} and μ_{21}.

20.8   As a result of the foregoing remarks, it is possible to assume that $\mathfrak{X}$ has at least one hyper element. (See 20.3.) *The discussion will now begin anew under this added assumption.*

Consider the 2-sphere hyper elements, $\mathfrak{S}_{11}, \cdots, \mathfrak{S}_{1a},$ of $\mathfrak{D}_1$ and $\mathfrak{S}_{21}, \cdots, \mathfrak{S}_{2a}$ of $\mathfrak{D}_2$. Recall that $\mathfrak{S}_{ik}$ contains an arc, $\bar{a}_{ik},$ and $\psi_i(\bar{a}_k) = \bar{a}_{ik}$, $i = 1, 2; k = 1, \cdots, a.$ (See (4) and (5).)

Let $\mathring{\delta}_1$ denote the collection of 2-spheres, $\mathfrak{S}_{11}, \cdots, \mathfrak{S}_{1a}.$ Notice that if $\mathfrak{S}_1 \in \mathring{\delta}_1$ and $\mathfrak{L}_1$ is a hyper element of $\mathfrak{X}_1,$ then $\mathfrak{L}_1 \cap \mathfrak{S}_1$ is empty or degenerate. (See (1, iv).)

A 2-sphere, $\mathfrak{S}_1 \in \mathring{\delta}_1,$ is said to be *orientably* [*non-orientably*] *connected* to a hyper element, $\mathfrak{L}_1,$ of $\mathfrak{X}_1$ if and only if $0 \neq \mathfrak{S}_1 \cap \mathfrak{L}_1 = a_{11}$ and there is a simple neighborhood [no simple neighborhood], $\mathfrak{N}_1,$ of a_{11} in $\mathfrak{X}_1$ (not in $\mathfrak{X}_1^*$) such that N_1 is orientable where $N_1 = m_{10}^{-1}(\mathfrak{N}_1).$ (See (8).) *Notice that $N_1 \approx N_2$ where $N_2 = \mu_{20}^{-1}(\mathfrak{N}_1)$ since m_{10} and μ_{20} are comparable.* (If $0 \neq \mathfrak{S}_1 \cap \mathfrak{L}_1,$ then $\mathfrak{S}_1$ is either orientably or non-orientably connected to $\mathfrak{L}_1.$)

Two cases now arise:

I. If $\mathfrak{S}_1 \in \mathring{\delta}_1$ and $\mathfrak{L}_1$ is a hyper element of $\mathfrak{X}_1$ such that $\mathfrak{S}_1 \cap \mathfrak{L}_1 \neq 0,$ then $\mathfrak{S}_1$ is non-orientably connected to $\mathfrak{L}_1.$

II. There exists an $\mathfrak{S}_1 \in \mathring{\delta}_1$ and a hyper element, $\mathfrak{L}_1,$ of $\mathfrak{X}_1$ to which $\mathfrak{S}_1$ is orientably connected.

20.9   *Case I.* Select a 2-sphere in $\mathring{\delta}_1$ *having a point in common with some hyper element of* $\mathfrak{X}_1$. It may be assumed that the 2-sphere is $\mathfrak{S}_{11}$.

Let $\mathfrak{X}_i^*$, $m_{i1}\colon X \rightrightarrows \mathfrak{X}_i^*$ and $\rho_i\colon \mathfrak{X}_i^* \rightrightarrows \mathfrak{X}_i$ be defined by formulas (9), (10) and (11), $i = 1, 2$.

The mapping γ_0 carries the subspace $\mathfrak{X}_2$ of $\mathfrak{X}_2^*$ homeomorphically onto the subspace $\mathfrak{X}_1$ of $\mathfrak{X}_1^*$; in particular, it maps the arc $\bar{a}_{21}$ in $\mathfrak{S}_{21}$ homeomorphically onto the arc $\bar{a}_{11}$ of $\mathfrak{S}_{11}$, and hence it can readily be extended to a homeomorphism, $\gamma_1\colon \mathfrak{X}_2^* \approx \mathfrak{X}_1^*$. Define:

$$(14) \qquad\qquad \mu_{21} = \gamma_1 m_{21}.$$

It is easy to see that m_{11} is *admissible* with respect to X_1, H_1, $\mathfrak{X}_1^*$ and $(\psi_1(\mathfrak{H}) \cup \mathfrak{S}_{11})$ while μ_2 is *admissible* with respect to X_2, H_2, $\mathfrak{X}_1^*$ and $(\psi_1(\mathfrak{H}) \cup \mathfrak{S}_{11})$.

Assertion: m_{11} *and* μ_{21} *are compatible.*

There is no difficulty in following the pattern of 20.4–20.7 to show that if $\mathfrak{U}_1$ is a normal region of $\mathfrak{X}_1^*$, then $U_1 \equiv m_{11}^{-1}(\mathfrak{U}_1) \approx \mu_{21}^{-1}(\mathfrak{U}_1) \equiv U_2$.

If $\mathfrak{U}_1$ is a *simple* region of $\mathfrak{X}_1^*$, U_1 is orientable, and x is an end point of some a_{1k} on a hyper element, $\mathfrak{L}_1$, of $\mathfrak{X}_1$, then $x \notin \mathfrak{U}_1$. For if $x \in \mathfrak{U}_1$, then by 12.6 there is a simple neighborhood, $\mathfrak{N}_1$, of x in $\mathfrak{X}_1$ such that $N_1 \equiv m_{10}^{-1}(\mathfrak{N}_1) \subset U_1$. Hence N_1 is orientable by 9.3, and $\mathfrak{S}_{1k}$ is orientably connected to $\mathfrak{L}_1$. This is impossible in *Case I*.

Consequently, since $\mathfrak{U}_1$ is connected, either $\mathfrak{U}_1 \subset \bigcup \mathfrak{L}_1$ or $\mathfrak{U}_1 \subset \mathfrak{X}_1 - \bigcup \mathfrak{L}_1$ where the unions are taken over the hyper elements, $\mathfrak{L}_1$, of $\mathfrak{X}_1$.

In the first instance the consistency requirement follows from the fact that m_{11} and m_{10} are identical on U_1, while μ_{21} and μ_{20} are identical on U_2 and m_{10} and μ_{20} are consistent. In the event that $\mathfrak{U}_1 \subset \mathfrak{X}_1 - \bigcup \mathfrak{L}_1$ but $\mathfrak{U}_1 \cap \mathfrak{S}_{11} = 0$, consistency is vacuously fulfilled. If $\mathfrak{U}_1 \subset \mathfrak{X}_1 - \bigcup \mathfrak{L}_1$ but $\mathfrak{U}_1 \cap \mathfrak{S}_{11} \neq 0$, then $\bar{\mathfrak{U}}_1 \cap \mathfrak{S}_{11}$ is the only cyclic element of $\bar{\mathfrak{U}}_1$ and the question of consistency offers no difficulty. (Recall that $\mathfrak{S}_{11} \cap \bigcup \mathfrak{L}_1 \neq 0$.)

20.10   *Case II.* It may be assumed that $\mathfrak{S}_{11}$ is orientably connected to $\mathfrak{L}_1$, a hyper element of $\mathfrak{X}_1$.

Define $\mathfrak{X}_1^*$, m_{i1} and ρ_i by formulas (9), (10) and (11), $i = 1, 2$. Further care must be exercised in the definition of $\gamma_1\colon \mathfrak{X}_2 \approx \mathfrak{X}_1$, a homeomorphic extension of $\gamma_0 : \mathfrak{X}_2 \approx \mathfrak{X}_1$. (Cf. *Case I.*)

Let $a_{11} = \mathfrak{S}_{11} \cap \mathfrak{L}_1$. Since $\mathfrak{S}_{11}$ is orientably connected to $\mathfrak{L}_1$, there is a sequence, $\{\mathfrak{N}_1^n\}$, of simple neighborhoods, $\mathfrak{N}_1^n$, of a_{11} in $\mathfrak{X}_1$ such that: $\mathfrak{N}_1^n \supset \mathfrak{N}_1^{n+1}$; $d(\mathfrak{N}_1^n) \to 0$; while if $N_1^n = m_{10}^{-1}(\mathfrak{N}_1^n)$ and $N_2^n = \mu_{20}^{-1}(\mathfrak{N}_1^n)$, then N_1^n and N_2^n are orientable, $n = 1, 2, 3, \cdots$ (see 12.6). The Peano space $\overline{\mathfrak{N}}_{1n}$ will have at least one 2-cell true cyclic element which is a subset of $\mathfrak{L}_1$, and $\mathfrak{N}_{1n} \cap a_{11}$ is an open subarc, β_1^n, of a_{11} with a_{11} for one of its end points, $n = 1, 2, 3, \cdots$.

Since m_{10} and μ_{20} are consistent and $\mathfrak{N}_1^n$ has at least one 2-cell true cyclic element, the consistency requirement determines *a unique matching iso-morphism*, $\kappa_n : H^2(N_2^n) \approx H^2(N_1^n)$, $n = 1, 2, 3, \cdots$ (see 14.1). *The isomorphism, κ_n, is fixed in the remainder of the argument,* $n = 1, 2, 3, \cdots$.

It is easy to see that one has commutativity in the following diagram of isomorphisms onto for $n = 1, 2, 3, \cdots$:

$$(15) \qquad \begin{array}{ccc} H^2(N_2^n) & \xleftarrow{\quad c_{2n} \quad} & H^2(N_2^{n+1}) \\[4pt] \kappa_n \downarrow & & \downarrow \kappa_{n+1} \\[4pt] H^2(N_1^n) & \xleftarrow{\quad c_{1n} \quad} & H^2(N_1^{n+1}) \end{array}$$

Notice that if $\mathfrak{N}_2^n = \gamma_0^{-1}(\mathfrak{N}_1^n)$, then $\mathfrak{N}_2^n \cap a_{21}$ is an open subarc, β_2^n, of a_{21} with $a_{21} \equiv \gamma_0^{-1}(a_{11}) = \mathfrak{S}_{21} \cap \mathfrak{L}_2$ for one of its end points, where $\mathfrak{L}_2 = \gamma_0^{-1}(\mathfrak{L}_1)$, and $\gamma_0(\beta_2^n) = \beta_1^n$, $n = 1, 2, 3, \cdots$. Consider a 2-cell in the open cylinder $\rho_i^{-1}(\beta_i^n)$ on $\mathfrak{S}_{i1}$, and let $\mathfrak{R}_i^n$ be its interior, $i = 1, 2; n = 1, 2, 3, \cdots$. If $K_i^n = m_{i1}^{-1}(\mathfrak{R}_i^n)$, then K_i^n is an open 2-cell in N_i^n since m_{i1}^{-1} is a homeomorphism on $\mathfrak{S}_{i1}^\circ$, $i = 1, 2; n = 1, 2, 3, \cdots$. Let $m_{i1n} = m_{i1}|K_i^n$ and consider the following diagram for $i = 1, 2; n = 1, 2, 3, \cdots$:

$$H^2(N_i^n) \xleftarrow{\quad a_{in} \quad} H^2(K_i^n) \xleftarrow{\quad m_{i1n}^* \quad} H^2(\mathfrak{R}_i^n) \xrightarrow{\quad b_{in} \quad} H^2(\mathfrak{S}_{i1})$$

All the homomorphisms are isomorphisms onto.

There is a homeomorphism, $\overline{\gamma} : \mathfrak{S}_{21} \approx \mathfrak{S}_{11}$ which agrees with γ_0 on $\overline{a}_{21}$ and, in addition, is such that the induced isomorphism, $\overline{\gamma}^* : H^2(\mathfrak{S}_{11}) \approx H^2(\mathfrak{S}_{21})$, has the following property:

$$\overline{\gamma}^* b_{11}(m_{111}^*)^{-1} a_{11}^{-1} \kappa_1 = b_{21}(m_{211}^*)^{-1} a_{21}^{-1}.$$

It follows from (15) that, for $n = 1, 2, 3, \cdots$:

$$(16) \qquad \bar{\gamma}^* b_{1n}(m_{11n}^*)^{-1} a_{1n}^{-1} \kappa_n = b_{2n}(m_{21n}^*)^{-1} a_{2n}^{-1}.$$

Define:

$$(17) \qquad \gamma_1(x_2) = \begin{cases} \gamma_0(x_2) & \text{if } x_2 \in X_2. \\[2ex] \bar{\gamma}(x_2) & \text{if } x_2 \in \mathfrak{S}_{21}. \end{cases}$$

The mapping γ_1 is a homeomorphism from X_2^* onto X_1^*. Let

$$\mu_{21} = \gamma_1 m_{21}.$$

20.11   There is no difficulty in showing that m_{11} is admissible with respect to X_1, H_1, X_1^* and $(\psi_1(\mathfrak{H}) \cup \mathfrak{S}_{11})$, while μ_{21} is admissible with respect to X_2, H_2, X_1^* and $(\psi_1(\mathfrak{H}) \cup \mathfrak{S}_{11})$.

20.12   It is a little more difficult to show that m_{11} and μ_{21} are compatible insofar as consistency is concerned.

Suppose that $\mathfrak{U}_1$ is a normal region of X_1^*. Let $U_1 = m_{11}^{-1}(\mathfrak{U}_1)$ $U_2 = \mu_{21}^{-1}(\mathfrak{U}_1)$.

The proof that $U_1 \approx U_2$ *is that of* 20.4–20.7.

From here on to 20.17, suppose $\mathfrak{U}_1$ is a *simple* region of X_1^*, while U_1 and U_2 are *orientable*. The argument will be in two subcases (cf. 20.3):

$$\text{Subcase } A. \quad \mathfrak{U}_1 \supset \mathfrak{S}_{11}.$$
$$\text{Subcase } B. \quad \mathfrak{U}_1 \not\supset \mathfrak{S}_{11}.$$

20.13   *Subcase A.* If $\mathfrak{U}_1^* = \rho_1(\mathfrak{U}_1)$, then $\mathfrak{U}_1^*$ is a simple region of X_1 while $\bar{\mathfrak{u}}_1$ and $\bar{\mathfrak{u}}_1^*$ have the same 2-cell true cyclic elements. Moreover, $m_{10}^{-1}(\mathfrak{U}_1^*) = m_{11}^{-1}(\mathfrak{U}_1) = U_1$ and $\mu_{20}^{-1}(\mathfrak{U}_1^*) = \mu_{21}^{-1}(\mathfrak{U}_1) = U_2$. Since m_{10} and μ_{20} are consistent, there is a matching isomorphism, $\eta: H^2(U_2) \approx H^2(U_1)$, determined by the consistency requirement. If $\bar{\mathfrak{u}}_1$ has no 2-cell true cyclic elements, then the consistency requirement for m_{21} and μ_{21} is vacuously fulfilled on $\mathfrak{U}_1$ with respect to the true cyclic elements of $\bar{\mathfrak{u}}_1$. Suppose, therefore, that $\bar{\mathfrak{u}}_1$ has the 2-cell true cyclic element $\mathfrak{E}$. Let $r: \bar{\mathfrak{u}}_1 \rightrightarrows \mathfrak{E}$ and $\bar{r}: \bar{\mathfrak{u}}_1^* \rightrightarrows \mathfrak{E}$ be the monotone retractions while $\underline{m}_{10} = m_{10}|U_1$, $\underline{m}_{11} = m_{11}|U_1$, $\underline{\mu}_{20} = \mu_{20}|U_2$, and

$\mu_{21} = \mu_{21}|U_2$. Now $\bar{r}(\mathfrak{U}_1^*) = r(\mathfrak{U}_1)$ since $\mathfrak{S} \cap \mathfrak{S}_{11}$ is empty or degenerate. Moreover, $\bar{r}m_{10} = rm_{11}$ and $\bar{r}\mu_{20} = r\mu_{21}$. Hence commutativity holds in the following diagram:

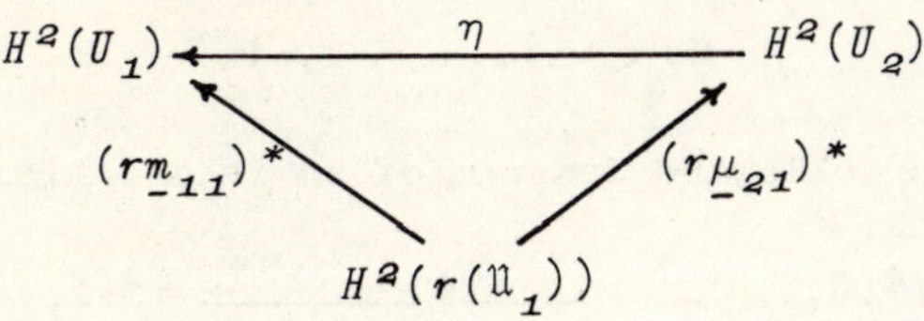

showing that m_{11} and μ_{21} are consistent on $\mathfrak{U}_1$ with respect to the 2-cell true cyclic elements of $\bar{\mathfrak{u}}_1$.

20.14   *Subcase B.* Here $\mathfrak{U}_1 \not\supset \mathfrak{S}_{11}$, and if $\mathfrak{U}_1 \cap \mathfrak{S}_{11} = 0$ then $\mathfrak{U}_1$ is a simple region of $\mathfrak{X}_1$. Hence the consistency of m_{11} and μ_{21} on $\mathfrak{U}_1$, with respect to the true cyclic elements of $\bar{\mathfrak{u}}_1$, follows as a direct consequence of the consistency of m_{10} and μ_{20}.

Suppose, therefore, that $\mathfrak{U}_1 \cap \mathfrak{S}_{11} \neq 0$. Compare the situation with 20.5 and recall that $\bar{\mathfrak{u}}_1 \not\supset \mathfrak{S}_{11}$. Suppose a_{11} and a_{12} are the end points of α_{11}, where $a_{11} = \mathfrak{S}_{11} \cap \mathfrak{L}_1$.

Four situations can now arise:

(i) $\mathfrak{U}_1 \cap (a_{11} \cup a_{12}) = 0$.

(ii) $\mathfrak{U}_1 \cap (a_{11} \cup a_{12}) = a_{11}$.

(iii) $\mathfrak{U}_1 \cap (a_{11} \cup a_{12}) = a_{12}$.

(iv) $\mathfrak{U}_1 \cap (a_{11} \cup a_{12}) = a_{11} \cup a_{12}$.

20.15   (i) In this case the consistency of m_{11} and μ_{21} on $\mathfrak{U}_1$, with respect to the true cyclic elements of $\bar{\mathfrak{u}}_1$, is obvious, as in 20.6 (i).

(ii) *In this case the proof differs considerably from 20.6 (ii) due to the possible presence of 2-cell true cyclic elements of $\bar{\mathfrak{u}}_1$ outside $\mathfrak{S}_{11}$.*

Refer to 20.10 and select an n so that:

$$\rho_1^{-1}(\mathfrak{R}_1^n) \subset \mathfrak{U}_1,$$

$$\rho_2^{-1}(\mathfrak{R}_2^n) \subset \gamma_1^{-1}(\mathfrak{U}_1) \equiv \mathfrak{U}_2.$$

For convenience, let $\mathfrak{R}_i^n$ be designated by $\mathfrak{R}_i$, $\mathfrak{R}_i^n$ by $\mathfrak{R}_i$, N_i^n by N_i, K_i^n by K_i, a_{in} by a_i, b_{in} by b_i, m_{i1n}^* by $\bar{m}_{i1}^*$, $i = 1, 2$, and κ_n by κ. (See 20.10.)

The set $\bar{\mathfrak{U}}_1 \cap \mathfrak{S}_{11}$ is a 2-cell true cyclic element, $\mathfrak{S}$, of $\bar{\mathfrak{U}}_1$, and $a_{11} \in \mathfrak{S}^\circ$. It is possible to select a 2-cell $\mathfrak{S}_1$ in $\mathfrak{S}^\circ - a_{11}$ such that $\mathfrak{S}_1^\circ \supset \mathfrak{R}_1 \cup \gamma_1(\mathfrak{R}_2)$. Hence, if $\mathfrak{S}_2 = \gamma_1^{-1}(\mathfrak{S}_1)$, then:

$$(18) \qquad\qquad \mathfrak{R}_i \subset \mathfrak{S}_i^\circ, \quad i = 1, \ 2.$$

Consider the following diagram of an isomorphism onto for $i = 1, \ 2$:

$$H^2(U_i) \xleftarrow{\qquad e_i \qquad} H^2(N_i)$$

and define:

$$(19) \qquad\qquad \eta = e_1 \kappa e_2^{-1}.$$

Then:

$$\eta : H^2(U_2) \approx H^2(U_1).$$

If $\mathfrak{U}_1^* = (\mathfrak{U}_1 - \mathfrak{S}_{11}) \cup \mathfrak{R}_1$, then $\mathfrak{U}_1^*$ is a simple region in $\mathfrak{X}_1$, while if $U_1^* = m_{10}^{-1}(\mathfrak{U}_1^*)$ and $U_2^* = \mu_{20}^{-1}(\mathfrak{U}_1^*)$, then $U_i^* \subset U_i$ and hence U_i^* is orientable by 9.3, $i = 1, \ 2$.

Consider the following diagram of an isomorphism onto for $i = 1, \ 2$:

$$H^2(U_i^*) \xleftarrow{\qquad \bar{e}_i \qquad} H^2(N_i),$$

and define:

$$\bar{\eta} = \bar{e}_1 \kappa \bar{e}_2^{-1}.$$

Since m_{10} and μ_{20} are compatible, they are consistent on $\mathfrak{U}_1^*$ with respect to every true cyclic element of $\bar{\mathfrak{U}}_1^*$, and it is a simple exercise to show that $\bar{\eta}$, in view of the uniqueness of κ, is a matching isomorphism for m_{10} and μ_{20} insofar as the true cyclic elements of $\bar{\mathfrak{U}}_1^*$ are concerned. If $A_1 = m_{10}^{-1}(\mathfrak{U}_1^* - \mathfrak{S}_{11})$ and $A_2 = \mu_{20}^{-1}(\mathfrak{U}_1^* - \mathfrak{S}_{11})$, then $m_{10}|A_1 = m_{11}|A_1$ and $\mu_{20}|A_2 = \mu_{21}|A_2$. Hence it follows that η is a matching isomorphism for m_{11} and μ_{21} insofar as the true cyclic elements of $\bar{\mathfrak{U}}_1$ — *with the possible exception of* $\mathfrak{S}_1$ — are concerned.

This latter case will be considered in more detail.

Let C_i be the 2-cell $m_{11}^{-1}(\mathfrak{S}_i)$, $\underline{m}_{11} = m_{11}|C_i^\circ$, $i = 1, \ 2$, and referring to (18), consider the following diagram for $i = 1, \ 2$:

$$H^2(N_i) \xleftarrow{\;a_i\;} H^2(K_i) \xleftarrow{\;\bar{m}^*_{i1}\;} H^2(\mathfrak{R}_i)$$

(diagram with vertical maps e_i, c_i, $\mathfrak{b}_i$ and diagonal maps $\mathfrak{b}_i$, $\mathfrak{g}_i$ into $H^2(\mathfrak{S}_{i1})$; bottom row)

$$H^2(U_i) \xleftarrow{\;\mathfrak{f}_i\;} H^2(C_i^o) \xleftarrow{\;\underline{m}^*_{i1}\;} H^2(\mathfrak{C}_i^o)$$

All the homomorphisms are isomorphisms onto, and commutativity holds in every box.

Recall that $\bar{\gamma} = \gamma_1 | \mathfrak{S}_{21}$. Let $\underline{\gamma}_1 = \gamma_1 | \mathfrak{C}_2^o$ and consider the following diagram in which commutativity holds and the homomorphisms are isomorphisms onto:

$$H^2(\mathfrak{S}_{11}) \xrightarrow{\;\bar{\gamma}^*\;} H^2(\mathfrak{S}_{21})$$

(vertical maps $\mathfrak{g}_1$ and $\mathfrak{g}_2$ upward)

$$H^2(\mathfrak{C}_1^o) \xrightarrow{\;\underline{\gamma}_1^*\;} H^2(\mathfrak{C}_2^o)$$

In view of (16), it is true that:

$$\bar{\gamma}^* \mathfrak{b}_1 (\bar{m}^*_{11})^{-1} a_1^{-1} \kappa = \mathfrak{b}_2 (\bar{m}^*_{21})^{-1} a_2^{-1}.$$

Therefore,

$$\bar{\gamma}^* \mathfrak{g}_1 (\underline{m}^*_{11})^{-1} \mathfrak{f}_1^{-1} e_1 \kappa = \mathfrak{g}_2 (\underline{m}^*_{21})^{-1} \mathfrak{f}_2^{-1} e_2.$$

Whence,

$$\begin{aligned}
\mathfrak{f}_1 \underline{m}^*_{11} &= [e_1 \kappa e_2^{-1}] \mathfrak{f}_2 \underline{m}^*_{21} [\mathfrak{g}_2^{-1} \bar{\gamma}^* \mathfrak{g}_1] \\
&= \eta \mathfrak{f}_2 \underline{m}^*_{21} \underline{\gamma}_1^* \quad \text{by (19)} \\
&= \eta \mathfrak{f}_2 (\underline{\gamma}_1 \underline{m}_{21})^*
\end{aligned}$$

and the proof for the consistency of m_{11} and μ_{21} on $\mathfrak{U}_{1'}$ with respect to every true cyclic element of $\bar{\mathfrak{u}}_{1'}$ is complete on observing that $\mu_{21} | C_2^o = \underline{\gamma}_1 \underline{m}_{21}$ and referring to 14.3.

<u>20.16</u>   (iii) Here $\mathfrak{U}_1 \cap (a_{11} \cup a_{12}) = a_{12}$ and $\mathfrak{U}_1 \not\supset \mathfrak{L}_1$. Since $a_{11} = \mathfrak{S}_{11} \cap \mathfrak{L}_{1'}$ it follows from 2.12 that if $a_{11} \in \bar{\mathfrak{u}}_{1'}$ then a_{11} is a point on a Jordan curve component, $\mathfrak{J}$, of $\dot{\mathfrak{u}}_1$. Let $\mathfrak{C}$ be the 2-cell true cyclic element of $\bar{\mathfrak{u}}_1$ bounded by $\mathfrak{J}$. Since x_1^* is a finite mantoid, there is a 2-cell $\mathfrak{R}$ such that $\mathfrak{R} \subset \mathfrak{C}^o$ and $\mathfrak{C}^o - \mathfrak{R}^o$ contains no local cut points of x_1^*. If $\mathfrak{B}_1$

$= \mathfrak{U}_1 - (\mathfrak{C}^\circ - \mathfrak{R}^\circ)$, then $\mathfrak{B}_1$ is a simple region of $\mathfrak{X}_1^*$, and $a_{11} \notin \overline{\mathfrak{B}}_1$. In the event $a_{11} \notin \overline{\mathfrak{U}}_1$, let $\mathfrak{B}_1 = \mathfrak{U}_1$. If $V_1 = m_{11}^{-1}(\mathfrak{B}_1)$ and $V_2 = \mu_{21}^{-1}(\mathfrak{B}_1)$, then $V_i \subset U_i$ and hence V_i is orientable, $i = 1, 2$.

By 20.10, if the integer n is large enough, then it is true that $\rho_1^{-1}(\overline{\mathfrak{R}}_{1n}) \cap \overline{\mathfrak{B}}_1 = 0$. It follows that if $\mathfrak{W}_1 = \mathfrak{B}_1 \cup \mathfrak{C}_{11} \cup \mathfrak{R}_{1n}$, then $\mathfrak{W}_1$ is a simple region of $\mathfrak{X}_1^*$, while if $W_1 = m_{11}^{-1}(\mathfrak{W}_1)$ and $W_2 = \mu_{21}^{-1}(\mathfrak{W}_1)$, then W_1 and W_2 are orientable. (Recall that $\mathfrak{C}_{11}$ is orientably connected to $\mathfrak{U}_1$).

The proof now proceeds through the following steps:

The mappings m_{11} and μ_{21} are consistent on $\mathfrak{B}_1$ with respect to the 2–cell true cyclic elements of $\overline{\mathfrak{B}}_1$ by subcase A (see 20.13).

By the argument of 20.15 (ii), they are consistent on $\mathfrak{B}_1$ *with respect to* $\mathfrak{C}_{11}$ and the 2–cell true cyclic elements of $\overline{\mathfrak{B}}_1$.

In view of this fact, it follows that they are consistent on $\mathfrak{B}_1$ with respect to the 2–cell true cyclic elements of $\overline{\mathfrak{W}}_1$.

Finally, it can be then shown that m_{11} and μ_{21} are consistent on $\mathfrak{U}_1$ with respect to the 2–cell true cyclic elements of $\overline{\mathfrak{U}}_1$.

<u>20.17</u>   (iv) If $\mathfrak{U}_1 \cap (a_{11} \cup a_{12}) = a_{11} \cup a_{12}$ (cf. 20.7), then first suppose $\overline{\mathfrak{U}}_1 \cap \mathfrak{C}_{11}$ is connected. If $\mathfrak{C} = \overline{\mathfrak{U}}_1 \cap \mathfrak{C}_{11}$, then $\mathfrak{C}$ is a 2–cell true cyclic element of $\overline{\mathfrak{U}}_1$. Let $\mathfrak{B}_1 = \mathfrak{U}_1$, $\mathfrak{W}_1 = \mathfrak{B}_1 \cup \mathfrak{C}_{11}$, and use the argument of 20.16.

If $\overline{\mathfrak{U}}_1 \cap \mathfrak{C}_{11}$ is not connected it must have two components, $\mathfrak{C}_1$ and $\mathfrak{C}_2$, each a 2–cell true cyclic element of $\overline{\mathfrak{U}}_1$. It must be assumed that $a_{1i} \in \mathfrak{C}_i^\circ$, $i = 1, 2$.

Dealing with the point a_{11} as in 20.15 (ii), it follows that there is a matching isomorphism, $\eta : H^2(U_2) \approx H^2(U_1)$, for m_{11} and μ_{21} insofar as the 2–cell true cyclic elements of $\overline{\mathfrak{U}}_1$ — with the possible exception of $\mathfrak{C}_2$ — are concerned. But if η is a matching isomorphism satisfactory for $\mathfrak{C}_1$, it must also be satisfactory for $\mathfrak{C}_2$ by the argument of 20.7.

Hence m_{11} and μ_{21} are consistent on $\mathfrak{U}_1$ with respect to every 2–cell true cyclic element of $\overline{\mathfrak{U}}_1$.

<u>20.18</u>   The situation is now this: the mapping m_{11} is admissible with respect to X_1, H_1, $\mathfrak{X}_1^*$, and $\psi_1(\mathfrak{H}) \cup \mathfrak{C}_{11}$; the mapping μ_{21} is admissible with respect to X_2, H_2, $\mathfrak{X}_1^*$, and $\psi_1(\mathfrak{H}) \cup \mathfrak{C}_{11}$; and the mappings m_{11} and μ_{21} are

compatible. Hence the process of 20.8–20.17 may be repeated, and it may be assumed that $\mathfrak{S}_{12}$ now plays the role of $\mathfrak{S}_{11}$. Thus if $\mathfrak{X}_{i_2} = \mathfrak{X}_i^* \cup \mathfrak{S}_{i_2}$, $i = 1, 2$, then there is a homeomorphic extension, $\gamma_2 : \mathfrak{X}_{22} \approx \mathfrak{X}_{12}$, of $\gamma_1 : \mathfrak{X}_2^* \approx \mathfrak{X}_1^*$ such that if $\mu_{22} = \gamma_2 m_{22}$ (see 20.3) then the mapping m_{12} is admissible with respect to X_1, H_1, $\mathfrak{X}_{12}$ and $\psi_1(\mathfrak{H}) \cup \mathfrak{S}_{11} \cup \mathfrak{S}_{12}$; the mapping μ_{22} is admissible with respect to X_2, H_2, $\mathfrak{X}_{12}$ and $\psi_1(\mathfrak{H}) \cup \mathfrak{S}_{11} \cup \mathfrak{S}_{12}$; and the mappings m_{12} and μ_{22} are compatible. Continuing in this fashion, in a finite number of steps one obtains a homeomorphism, $\gamma : \mathfrak{Y}_2 \approx \mathfrak{Y}_1$ (which is an extension of $\gamma_0 : \mathfrak{X}_2 \approx \mathfrak{X}_1$) such that ϕ_1 and $\gamma \phi_2$ are compatible. Let $\mathfrak{Y} = \mathfrak{Y}_1$, $\mu_1 = \phi_1$, $\mu_2 = \gamma \phi_2$ and $\omega = m_1 \phi_1^{-1}$, then by 20.2 the proof of the theorem is complete.

20.19   THEOREM:

Given:

    1°. *A mapping, m_i, admissible with respect to X_i, H_i, $\mathfrak{X}$ and $\mathfrak{H}$.*

    2°. *$\mathfrak{X}$ is a finite mantoid.*

    3°. *m_1 and m_2 are o–compatible.*

    4°. *$\epsilon > 0$.*

Conclusion: *There is a cluster $\mathfrak{Y}$, a mapping $\omega : \mathfrak{Y} \rightrightarrows \mathfrak{X}$, and mappings μ_1 and μ_2 such that for $i = 1, 2$:*

    5°. *μ_i is admissible with respect to X_i, H_i, $\mathfrak{Y}$ and $\omega^{-1}(\mathfrak{H})$.*

    6°. *μ_1 and μ_2 are o–compatible.*

    7°. *$\overline{\rho}\{m_i, \omega \mu_i\} < \epsilon$.*

Proof: The proof follows that of theorem 20.1 through 20.2, but the extension of the homeomorphism $\gamma_0 : \mathfrak{X}_2 \approx \mathfrak{X}_1$ to obtain a homeomorphism $\gamma : \mathfrak{Y}_2 \approx \mathfrak{Y}_1$ is much simpler.

Since m_{10} and μ_{20} are o–consistent, there is a matching isomorphism, $\eta : H^2(X_2) \approx H^2(X_1)$. (If there are no hyper elements in $\mathfrak{X}_1$, then η is not uniquely determined, but this makes no difference in the argument.)

Suppose $\mathfrak{R}_{ik}$ is a 2-cell on $[\mathfrak{S}_{ik} - \overline{(\mathfrak{X}_i - \mathfrak{S}_{ik})}]$, $i = 1, 2$; $k = 1, \cdots, a$. Then $\phi_i^{-1}(\mathfrak{R}_{ik})$ is a 2-cell K_{ik} on X_i since ϕ_i^{-1} is a homeomorphism on $\mathfrak{R}_{ik}$, $i = 1, 2$; $k = 1, \cdots, a$. Let $\phi_{ik} = \phi_i | K_{ik}^\circ$ and consider the following diagram for $i = 1, 2$; $k = 1, \cdots, a$:

$$H^2(X_i) \xleftarrow{\quad a_{ik} \quad} H^2(K^{\circ}_{ik})$$

$$\uparrow \phi^*_{ik}$$

$$H^2(\mathfrak{S}_{ik}) \xleftarrow{\quad \mathfrak{b}_{ik} \quad} H^2(\mathfrak{R}^{\circ}_{ik})$$

There is a homeomorphism, $\gamma_k : \mathfrak{S}_{2k} \approx \mathfrak{S}_{1k}$, which agrees with γ_0 on $\bar{a}_{2k}$ and, in addition, is such that the induced isomorphism, $\gamma^*_k : H^2(\mathfrak{S}_{1k}) \approx H^2(\mathfrak{S}_{2k})$ has the following property:

$$\gamma^*_k \mathfrak{b}_{1k} (\phi^*_{1k})^{-1} a^{-1}_{1k} \eta = \mathfrak{b}_{2k} (\phi^*_{2k})^{-1} a^{-1}_{2k}.$$

(Cf. 20.10)

Define:

$$\gamma(\mathfrak{x}_2) = \begin{cases} \gamma_0(\mathfrak{x}_2) & \text{if } \mathfrak{x}_2 \in \mathfrak{X}_2. \\[2ex] \gamma_k(\mathfrak{x}_2) & \text{if } \mathfrak{x}_2 \in \mathfrak{S}_{2k}, \quad k = 1, \cdots, a. \end{cases}$$

Now γ is a homeomorphism from $\mathfrak{Y}_2$ onto $\mathfrak{Y}_1$ and if $\omega = \theta_1$, $\mu_1 = \phi_1$, $\mu_2 = \gamma\phi_2$, then it follows that μ_1 and μ_2 are o—comparable by an obvious modification of 20.6, while o—consistency is a simple exercise, in contradistinction to the situation in theorem 20.1. The discussion of 20.2 completes the argument.

21.   The fourth modification

The purpose of this section is to show that if the common image of the monotone mappings under consideration is a cluster, then, roughly speaking, it can be replaced by a 2—manifold homeomorphic to the range spaces of the mono—tone mappings.

21.1   THEOREM:

Given:

1°. *A mapping, m_i, admissible with respect to X_i, H_i, $\mathfrak{X}$ and $\mathfrak{H}$, $i = 1, 2$.*

2°. *$\mathfrak{X}$ is a cluster.*

3°. *m_1 and m_2 are compatible.*

4°. *$\epsilon > 0$.*

Conclusion: *There is a 2—manifold $\mathfrak{Y}$ homeomorphic to X_i, a map—*

ping $\omega:\mathcal{Y} \rightrightarrows \mathcal{X},$ *and a pair of mappings* μ_1 *and* μ_2 *such that for* $i = 1, 2:$

5°. μ_i *is admissible with respect to* $X_i,$ $H_i,$ $\mathcal{Y}$ *and* $\omega^{-1}(\mathfrak{H}).$

6°. $\bar{\rho}\{m_i, \omega\mu_i\} < \epsilon.$

Proof: Suppose that $\mathfrak{L}$ is a hyper element of $\mathcal{X},$ that $\mathfrak{T}$ is the true cyclic element of $\mathcal{X}$ containing $\mathfrak{L},$ that $\bar{r}:\mathcal{X} \rightrightarrows \mathfrak{T}$ is the monotone retraction, and that $\widetilde{m}_i = \bar{r}m_i,$ $i = 1, 2.$ There is a finite set, $\mathfrak{F}(\mathfrak{L}, m_i),$ consisting of precisely those points, $x,$ on $\mathfrak{L}$ for which $R(\widetilde{m}_i^{-1}(x)) \neq 0,$ $i = 1, 2$ (see 13.1). Let $\mathfrak{C}(\mathfrak{L})$ be the set of local cut points of $\mathcal{X}$ in $\mathfrak{L},$ and $\mathfrak{F}^*(\mathfrak{L}, m_i) = \mathfrak{F}(\mathfrak{L}, m_i) - \mathfrak{C}(\mathfrak{L}),$ $i = 1, 2.$

If $x \in \mathfrak{F}^*(\mathfrak{L}, m_1),$ then $m_1^{-1}(x) = \widetilde{m}_1^{-1}(x),$ and $R(m_1^{-1}(x)) \neq 0.$ If $R(m_2^{-1}(x)) = 0,$ then $x \in \mathfrak{L} - \mathfrak{F}(\mathfrak{L}, m_2).$ Suppose G is any open set containing $m_1^{-1}(x).$ There is a 2-cell, $\mathfrak{C},$ on $\mathfrak{L}$ such that:

(i) $x \in \mathfrak{C}^\circ \subset \mathfrak{L} - \mathfrak{F}(\mathfrak{L}, m_2).$

(ii) $\mathfrak{C}$ contains no local cut point of $\mathcal{X}.$

(iii) $m_1^{-1}(\mathfrak{C}) \subset G.$

Now $\widetilde{m}_2^{-1}(\mathfrak{C}^\circ) = m_2^{-1}(\mathfrak{C}^\circ),$ but $\widetilde{m}_2^{-1}(\mathfrak{C}^\circ)$ is an open 2-cell by 13.1. Since m_1 and m_2 are comparable, $m_1^{-1}(\mathfrak{C}^\circ)$ is also an open 2-cell. Hence $R(m_1^{-1}(x)) = 0.$ This contradiction serves to show that if $x \in \mathfrak{F}^*(\mathfrak{L}, m_1),$ then $x \in \mathfrak{F}^*(\mathfrak{L}, m_2).$ The reverse inclusion follows by a similar reasoning, therefore $\mathfrak{F}^*(\mathfrak{L}, m_1) = \mathfrak{F}^*(\mathfrak{L}, m_2) \equiv \mathfrak{F}^*(\mathfrak{L}).$

If $x \in \mathfrak{C}(\mathfrak{L}),$ then $R(m_i^{-1}(x)) \neq 0$ by the argument used in 13.1, $i = 1, 2.$ Hence if $x \in \mathfrak{L},$ then $R(m_i^{-1}(x)) \neq 0$ if and only if $x \in \mathfrak{F}^*(\mathfrak{L}) \cup \mathfrak{C}(\mathfrak{L}) \equiv \mathfrak{F}(\mathfrak{L}),$ $i = 1, 2.$ (This shows that $\mathfrak{F}(\mathfrak{L}, m_1) = \mathfrak{F}(\mathfrak{L}, m_2).$)

<u>21.2</u>   Notice that $\mathfrak{F}(\mathfrak{L})$ is a finite set, and define $\mathfrak{F}$ to be $\bigcup \mathfrak{F}(\mathfrak{L})$ where $\mathfrak{L}$ ranges over the class of hyper elements of $\mathcal{X}.$ Since $\mathcal{X}$ is a cluster, the set $\mathfrak{F}$ is finite; suppose it consists of the points $q_1, \cdots, q_q.$

It is easy to see that there is a system of simple neighborhoods, $\mathfrak{U}_1, \cdots, \mathfrak{U}_q,$ such that:

(1)   (i) $q_k \in \mathfrak{U}_k,$ $k = 1, \cdots, q.$

(ii) $\bar{\mathfrak{U}}_k \cap m_i(\overline{X_i - H_i}) = 0,$ $k = 1, \cdots, q;$ $i = 1, 2.$

(iii) $\bar{\mathfrak{U}}_j \cap \bar{\mathfrak{U}}_k = 0,$ $j \neq k;$ $j, k = 1, \cdots, q.$

(iv) $d(\mathfrak{U}_k) < \epsilon,$ $k = 1, \cdots, q.$

(Property (ii), for example, can be obtained since m_i is admissible with respect to X_i, H_i, $\mathfrak{X}$ and $\mathfrak{H}$. Hence if $x_i \in \overline{X_i - H_i}$, then $x_i = m_i^{-1} m_i(x_i)$, and therefore $R(m_i^{-1} m_i(x_i)) = 0$, $i = 1, 2$.)

Let $\mathfrak{B} = \mathfrak{X} - \bigcup \mathfrak{U}_k$ and $B_i = m_i^{-1}(\mathfrak{B})$, $i = 1, 2$. Suppose that $|\mathfrak{Y}_i|$ is the completion, over X_i, of the upper semi-continuous decomposition of B_i determined by $m_i | B_i$, and that $\phi_i : X \rightrightarrows \mathfrak{Y}_i$ is the associated mapping, $i = 1$, 2 (cf. 20.1 and see Youngs [12]). If $\mathfrak{y}_i \in \mathfrak{Y}_i$, then it is easy to see that $R(\phi_i^{-1}(\mathfrak{y}_i)) = 0$; hence $\mathfrak{Y}_i \approx X_i$ by 6.5, $i = 1, 2$.

If, for $i = 1, 2$,

$$(2) \qquad\qquad \mathfrak{B}_i = \phi_i(B_i)$$

and

$$(3) \qquad\qquad \theta_i = m_i \phi_i^{-1},$$

then $\theta_i : \mathfrak{Y}_i \rightrightarrows \mathfrak{X}$ is monotone, and $\theta_i^{-1} | \mathfrak{B}$ is a homeomorphism from $\mathfrak{B}$ onto $\mathfrak{B}_i$. If $\mathfrak{U}_{ik} = \theta_i^{-1}(\mathfrak{U}_k)$, $k = 1, \cdots, q$; $i = 1, 2$, then the components of $\mathfrak{Y}_i - \mathfrak{B}_i$ are $\mathfrak{U}_{i1}, \cdots, \mathfrak{U}_{iq}$, $i = 1, 2$. If the components of $\mathring{\mathfrak{U}}_k$ are $\mathfrak{I}_{k1}, \cdots, \mathfrak{I}_{ka_k}$, $k = 1, \cdots, q$, then these are Jordan curves disjoint in pairs. Hence $\mathring{\mathfrak{U}}_{ik}$ has components $\mathfrak{I}_{ik1} = \theta_i^{-1}(\mathfrak{I}_{k1}), \cdots, \mathfrak{I}_{ika_k} = \theta_i^{-1}(\mathfrak{I}_{ka_k})$, $k = 1, \cdots, q$; $i = 1, 2$, and these are Jordan curves disjoint in pairs. It follows that $\overline{\mathfrak{U}}_{ik}$ is a 2-manifold bounded by $\mathfrak{I}_{ik1}, \cdots, \mathfrak{I}_{ika_k}$, $k = 1, \cdots, q$; $i = 1, 2$.

If $\underline{\theta}_i = \theta_i | \mathfrak{B}_i$, $i = 1, 2$, let

$$(4) \qquad\qquad \gamma_0 = \underline{\theta}_1^{-1} \underline{\theta}_2.$$

Now γ_0 is a homeomorphism from $\mathfrak{B}_2$ onto $\mathfrak{B}_1$ which carries the Jordan curve $\mathfrak{I}_{2kb}$ in $\mathring{\mathfrak{U}}_{2k}$ topologically onto the Jordan curve $\mathfrak{I}_{1kb}$ in $\mathring{\mathfrak{U}}_{1k}$, $k = 1, \cdots, q$; $b = 1, \cdots, a_k$. It will be shown that γ_0 can be extended to a homeomorphism, $\gamma : \mathfrak{Y}_2 \approx \mathfrak{Y}_1$. In view of the preceding remarks, this will be the case if and only if γ_0 can be extended so as to map $\overline{\mathfrak{U}}_{2k}$ topologically onto $\overline{\mathfrak{U}}_{1k}$, $k = 1, \cdots, q$.

For notational convenience, suppose that $\mathfrak{U}$ is any one of the sets $\mathfrak{U}_1, \cdots, \mathfrak{U}_q$, that q is the element of $\mathfrak{F}$ in $\mathfrak{U}$, that $\theta_i^{-1}(\mathfrak{U}) = \mathfrak{U}_i$, $i = 1, 2$, that the components of $\mathring{\mathfrak{U}}$ are $\mathfrak{I}_1, \cdots, \mathfrak{I}_a$, and that $\mathfrak{I}_{ib} = \theta_i^{-1}(\mathfrak{I}_b)$, $b = 1, \cdots, a$; $i = 1, 2$. Notice that $\mathfrak{U}_i \approx m_i^{-1}(\mathfrak{U}) \equiv U_i$, $i = 1, 2$. Hence in view of the comparability of m_1 and m_2, one has $\overline{\mathfrak{U}}_1 \approx \overline{\mathfrak{U}}_2$. Therefore, in view of 10.9, the homeomorphism γ_0 can certainly be extended so as to map $\overline{\mathfrak{U}}_2$ topologically onto $\overline{\mathfrak{U}}_1$ *except possibly* in case $\overline{\mathfrak{U}}_1$ is orientable and $a > 1$.

To take care of this case, let $\bar{\phi}_i = \phi_i|\bar{U}_i$, $\bar{\theta}_i = \theta_i|\bar{\mathfrak{u}}_i$, $\bar{m}_i = m_i|\bar{U}_i$, $i = 1, 2$. Then $\bar{\phi}_i : \bar{U}_i \rightrightarrows \bar{\mathfrak{u}}_i$ and is one-to-one in U_i; moreover, $\bar{m}_i = \bar{\theta}_i\bar{\phi}_i$, $i = 1, 2$.

Since $\mathfrak{u}$ is a simple neighborhood containing q and such that $\mathfrak{u} \cap \mathfrak{F} = q$, the true cyclic elements of $\bar{\mathfrak{u}}$ are 2-cells, $\mathfrak{K}_1, \cdots, \mathfrak{K}_a$, bounded by $\mathfrak{I}_1, \cdots, \mathfrak{I}_a$, respectively, and $\bar{\mathfrak{u}} = \bigcup \mathfrak{K}_b$. (See 12.6.)

<u>21.3</u>   Suppose $r_b : \bar{\mathfrak{u}} \rightrightarrows \mathfrak{K}_b$ is the monotone retraction, $b = 1, \cdots, a$. It is convenient to consider the compactifications $\bar{U}_1$, $\bar{U}_2$ and $\mathfrak{K}_b$ of U_1, U_2 and $r_b(\mathfrak{u})$, respectively. Now since m_1 and m_2 are consistent, there is a matching isomorphism, $\eta : H^2(\bar{U}_2, \dot{U}_2) \approx H^2(\bar{U}_1, \dot{U}_1)$ such that commutativity holds in the following diagram for $b = 1, \cdots, a$ (see 13.7 and 14.1):

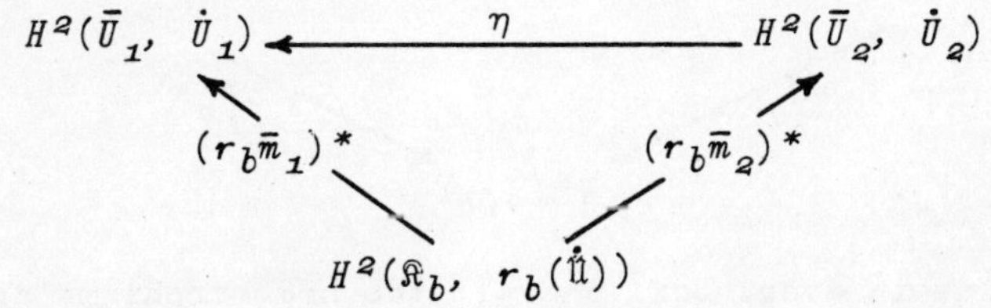

$$H^2(\bar{U}_1, \dot{U}_1) \xleftarrow{\quad\eta\quad} H^2(\bar{U}_2, \dot{U}_2)$$

$$(r_b\bar{m}_1)^* \qquad\qquad (r_b\bar{m}_2)^*$$

$$H^2(\mathfrak{K}_b, r_b(\dot{\mathfrak{u}}))$$

Now consider the following diagram for $b = 1, \cdots, a; i = 1, 2$:

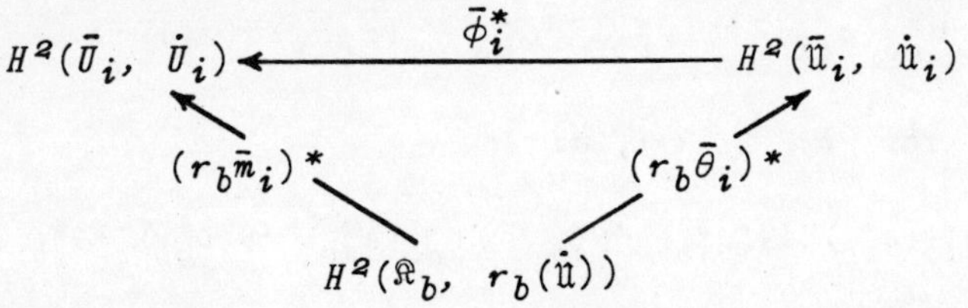

$$H^2(\bar{U}_i, \dot{U}_i) \xleftarrow{\quad\bar{\phi}_i^*\quad} H^2(\bar{\mathfrak{u}}_i, \dot{\mathfrak{u}}_i)$$

$$(r_b\bar{m}_i)^* \qquad\qquad (r_b\bar{\theta}_i)^*$$

$$H^2(\mathfrak{K}_b, r_b(\dot{\mathfrak{u}}))$$

Commutativity holds, and all the homomorphisms are isomorphisms onto. Hence if $\kappa = (\bar{\phi}_1^*)^{-1}\eta\bar{\phi}_2^*$, then $\kappa : H^2(\bar{\mathfrak{u}}_2, \dot{\mathfrak{u}}_2) \approx H^2(\bar{\mathfrak{u}}_1, \dot{\mathfrak{u}}_1)$ and

$$(5) \qquad\qquad (r_b\bar{\theta}_1)^* = \kappa(r_b\bar{\theta}_2)^*, \quad b = 1, \cdots, a.$$

Let $|\mathfrak{M}_{ib}|$ be the upper semi-continuous decomposition of $\bar{\mathfrak{u}}_i$ consisting of the sets $\mathfrak{I}_{i1}, \cdots, \hat{\mathfrak{I}}_{ib}, \cdots, \mathfrak{I}_{ia}$ and single points of $\mathfrak{u}_i \cup \mathfrak{I}_{ib}$. Suppose $\nu_{ib} : \bar{\mathfrak{u}}_i \rightrightarrows \mathfrak{M}_{ib}$ is the associated mapping, $b = 1, \cdots, a; i = 1, 2$ (cf. 9.9). For $b = 1, \cdots, a; i = 1, 2$, let

$$(6) \qquad\qquad \psi_{ib} = r_b\bar{\theta}_i\nu_{ib}^{-1}.$$

Then $\psi_{1b} : \mathfrak{M}_{ib} \rightrightarrows \mathfrak{K}_b$ is a monotone mapping, and for $b = 1, \cdots, a; i = 1, 2$:

$$(7) \qquad\qquad \psi_{ib}\nu_{ib} = r_b\bar{\theta}_i.$$

Let $\theta_{ib} = \bar{\theta}_i | \Im_{ib}$ and consider the following diagram for $b = 1,$ $\cdots, a;$ $i = 1, 2$:

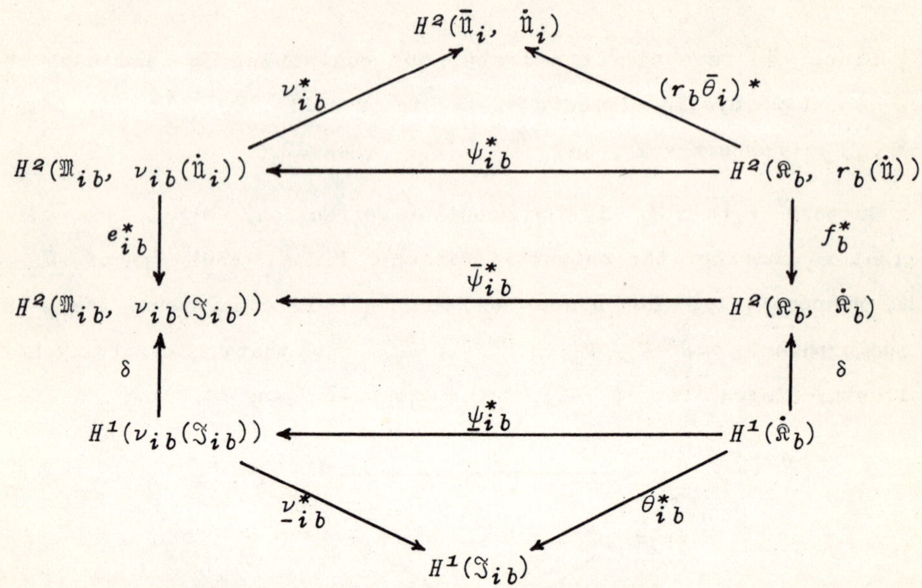

Commutativity holds in every box, and all the homomorphisms are isomorphisms onto. Hence for $i = 1, 2;$ $b = 1, \cdots, a$:

$$\nu_{ib}^{*}(e_{ib}^{*})^{-1}\delta(\underline{\nu}_{ib}^{*})^{-1}\theta_{ib}^{*} = (r_b\bar{\theta}_i)^{*}(f_b^{*})^{-1}\delta.$$

Therefore by (5), for $b = 1, \cdots, a$:

$$\nu_{1b}^{*}(e_{1b}^{*})^{-1}\delta(\underline{\nu}_{1b}^{*})^{-1}\theta_{1b}^{*} = \kappa\nu_{2b}^{*}(e_{2b}^{*})^{-1}\delta(\underline{\nu}_{2b}^{*})^{-1}\theta_{2b}^{*}.$$

Let $\dot{h} = \gamma_o | \dot{\mathfrak{u}}_2$ and $h_b = \dot{h} | \Im_{2b}$, then $h_b = \theta_{1b}^{-1}\theta_{2b}$, and in view of 9.9 one has commutativity in the following diagram of isomorphisms onto, for $b = 1, \cdots, a$:

$$
\begin{array}{ccc}
H^2(\bar{\mathfrak{u}}_1, \dot{\mathfrak{u}}_1) & \xleftarrow{\;\delta^*\;} & H^2(\Im_{1b}) \\[2mm]
\kappa\uparrow & & \downarrow h_b^* \\[2mm]
H^2(\bar{\mathfrak{u}}_2, \dot{\mathfrak{u}}_2) & \xleftarrow{\;\delta^*\;} & H^1(\Im_{2b})
\end{array}
$$

Consequently, $\dot{h}$ carries a concordant orientation of $\dot{\mathfrak{u}}_2$ into a concordant orientation of $\dot{\mathfrak{u}}_1$ and hence, by 10.9, can be extended to a homeomorphism $h : \bar{\mathfrak{u}}_2 \approx \bar{\mathfrak{u}}_1$.

This completes the proof that $\gamma_0 : \mathcal{B}_2 \approx \mathcal{B}_1$ can be extended to a homeomorphism, $\gamma : \mathcal{Y}_2 \approx \mathcal{Y}_1$.

Define $\mathcal{Y} = \mathcal{Y}_1$, $\omega = \theta_1$, $\mu_1 = \phi_1$ and $\mu_2 = \gamma\phi_2$. Now $\mathcal{Y} \approx X_i$, $i = 1$, 2, by 21.2, and the argument for items 5° and 6° is a precise duplication of 20.2.

21.4　　　　THEOREM:

Given:

1° *A mapping,* m_i, *admissible with respect to* X_i, H_i, X *and* $\mathfrak{H}$, $i = 1$, 2.

2° X *is a cluster.*

3° m_1 *and* m_2 *are o-compatible.*

4° $\epsilon > 0$.

Conclusion: *There is a 2-manifold* $\mathcal{Y}$ *homeomorphic to* X_i, *a mapping* $\omega : \mathcal{Y} \rightrightarrows X$, *and a pair of mappings* μ_1 *and* μ_2 *such that for* $i = 1$, 2:

5° μ_i *is admissible with respect to* X_i, H_i, $\mathcal{Y}$ *and* $\omega^{-1}(\mathfrak{H})$.

6° $\bar\rho\{m_i, \omega\mu_i\} < \epsilon$.

Proof: The argument is that of theorem 21.1 through 21.2. To use the remainder of the proof of theorem 21.1, it is only necessary to show that o–consistency implies the consistency of m_1 and m_2 on $\mathfrak{U}$ with respect to the true cyclic elements of $\bar{\mathfrak{U}}$ where $\mathfrak{U}$ is a simple neighborhood in the collection $\mathfrak{U}_1, \cdots, \mathfrak{U}_q$, and this is an easy exercise.

22.　The final modification and sufficiency theorem

22.1　　　　The modification theorem for this section is an immediate consequence of the approximation theorem of 10.5 and the subsequent remark.

THEOREM:

Given:

1° *A mapping,* m_i, *admissible with respect to* X_i, H_i, X *and* $\mathfrak{H}$, $i = 1$, 2.

2° $X \approx X_i$, $i = 1$, 2.

3° $\epsilon > 0$.

Conclusion:

4° *There is a homeomorphism,* h_i, *admissible with respect to*

X_i, H_i, $\mathfrak{X}$ and $\mathfrak{H}$, $i = 1$, 2.

$\quad\quad$ 5° $\bar{\rho}\{m_i, \mu_i\} < \epsilon$, $i = 1$, 2.

22.2 $\quad$ Using the chain of modification theorems which have been proved in sections 18—22, one readily obtains the following result:

THEOREM:

Given:

$\quad\quad$ 1° *A mapping,* m_i, *weakly admissible with respect to* X_i, H_i, $\mathfrak{X}$ *and* $\mathfrak{H}$, $i = 1$, 2.

$\quad\quad$ 2° m_1 *and* m_2 *are compatible [o—compatible].*

$\quad\quad$ 3° $\epsilon > 0$.

Conclusion:

$\quad\quad$ 4° *There is a 2—manifold* $\mathfrak{Y}$ *homeomorphic to* X_i, *a set* $\mathfrak{R}$ *in* $\mathfrak{Y}$, *and a mapping* $\omega:\mathfrak{Y} \to \mathfrak{X}$.

$\quad\quad$ 5° *There is a homeomorphism,* h_i, *admissible with respect to* X_i, H_i, $\mathfrak{Y}$ *and* $\mathfrak{R}$, $i = 1$, 2.

$\quad\quad$ 6° $\bar{\rho}\{m_i, \omega h_i\} < \eta$, $i = 1$, 2.

Remark: Of the five modifications which go into this final result, the third is certainly the most difficult to prove. Moreover, the proof of the fourth modification theorem goes through if the hypothesis is weakened to state that m_1 and m_2 are consistent on any *simple neighborhood* $\mathfrak{R}$ with respect to the true cyclic elements of $\bar{\mathfrak{R}}$. Consequently, the chain of modification theorems would prove theorem 22.2 if the conclusion of the third was weakened to state that μ_1 and μ_2 are consistent on any simple neighborhood $\mathfrak{R}$ with respect to the true cyclic elements of $\bar{\mathfrak{R}}$. One might expect that this weakened third modification theorem would be considerably simpler to prove. Unfortunately, this does not appear to be the case.

22.3 $\quad$ In considering theorem 22.2, if m_1 and m_2 are o—compatible then X_1 and X_2 are orientable 2—manifolds, and there is a matching isomorphism, $\eta:H^2(X_2) \approx H^2(X_1)$. An examination of the successive modifications shows that in respect to the isomorphism $h_i^*:H^2(\mathfrak{Y}) \approx H^2(X_i)$ induced by the homeomorphism $h_i:X_i \approx \mathfrak{Y}_i$ of the conclusion, $i = 1$, 2, one has $h_1^* = \eta h_2^*$. This item will be of importance at the end of the next chapter.

22.4   In view of 22.2 it is a simple matter to prove the principal sufficiency condition.

THEOREM:

Given:

1°. *A mapping, m_i, weakly admissible with respect to X_i, H_i,
$\mathfrak{X}$ and $\mathfrak{H}$, $i = 1, 2$.*

2°. *m_1 and m_2 are compatible [o-compatible].*

3°. *$\underline{m}_i = m_i \,|\, H_i$, $i = 1, 2$.*

Conclusion: $m_1 \sim m_2$ and $\underline{m}_1 \sim \underline{m}_2$.

Proof: Suppose $\epsilon > 0$. By theorem 22.2 there is a 2-manifold, $\mathfrak{Y}$,
homeomorphic to X_i, a set $\mathfrak{R}$ in $\mathfrak{Y}$, a mapping $\omega : \mathfrak{Y} \to \mathfrak{X}$, and homeomorphisms
h_1 and h_2 such that:

(i) h_i is admissible with respect to X_i, H_i, $\mathfrak{Y}$ and $\mathfrak{R}$,
$i = 1, 2$.

(ii) $\bar{\rho}\{m_i,\, \omega h_i\} < \epsilon/2$, $i = 1, 2$.

Define $h_\epsilon = h_2^{-1} h_1$. Now h_ϵ is a homeomorphism from X_1 onto X_2 mapping H_1
onto H_2. Moreover:

$$\bar{\rho}\{m_1,\, m_2 h_\epsilon\} \leq \bar{\rho}\{m_1,\, \omega h_1\} + \bar{\rho}\{\omega h_1,\, m_2 h_\epsilon\}$$

$$< \epsilon/2 + \bar{\rho}\{\omega h_2 h_\epsilon,\, m_2 h_\epsilon\}$$

$$< \epsilon$$

Hence $m_1 \sim m_2$, and as $h_\epsilon(H_1) = H_2$ it is also true that $\underline{m}_1 \sim \underline{m}_2$.

22.5   In view of the remarks in 22.3, if m_1 and m_2 are o-compatible
while $\eta : H^2(X_2) \approx H^2(X_1)$ is a matching isomorphism, then the matching homeomorphism h_ϵ of theorem 22.4 may be selected so that $\eta = h_\epsilon^*$.

CHAPTER VI

F - CRITERIA

The reduction theorem of chapter I together with the results of chapters IV and V furnish one with an immediate wealth of F–criteria. A few of these criteria are catalogued here. In addition, it is possible to introduce oriented Fréchet equivalence — an oft useful concept in analysis — and present an appropriate characterization.

23. Table of results

A convenient way of exhibiting some of the available F–criteria is by means of an implication diagram summarizing the results of the paper. A vertical arrow in the diagrams below means that the statement at the beginning of the arrow implies the statement at the end.

__23.1__   Given two mappings, $f_1:X_1 \to Y$ and $f_2:X_2 \to Y$, where X_1 and X_2 are closed 2–manifolds:

$$f_1 \sim f_2$$

$$\downarrow \quad \text{(by 1.8)}$$

There is a monotone–light factorization, lm_i, of f_i with middle space X, $i = 1, 2$, such that $m_1 \sim m_2$.

$$\downarrow \quad \text{(by 16.1)}$$

There is a monotone–light factorization, lm_i, of f_i with middle space X, $i = 1, 2$, such that if U is any normal region in X while $U_i = m_i^{-1}(U)$ and $\underline{m}_i = m_i | U_i$, $i = 1, 2$, then there is a homeomorphism, $h:U_1 \approx U_2$, guaranteeing commutativity in the following diagram:

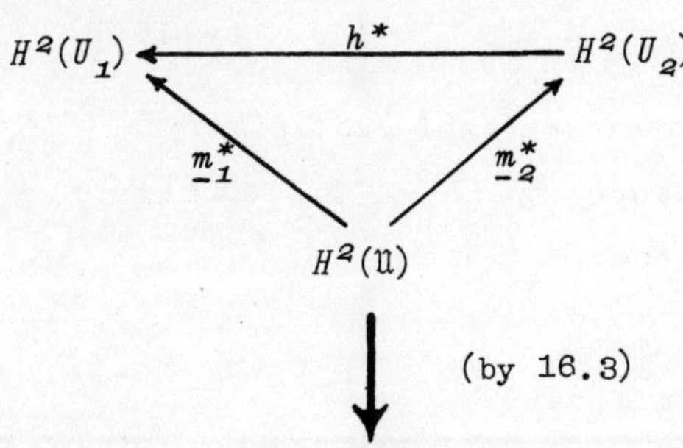

(by 16.3)

There is a monotone–light factorization, lm_i, of f_i with middle space $\mathfrak{X}$, $i = 1, 2$, such that m_1 and m_2 are compatible.

(by 22.4)

There is a monotone–light factorization, lm_i, of f_i with middle space $\mathfrak{X}$, $i = 1, 2$, such that $m_1 \sim m_2$.

(by 1.8)

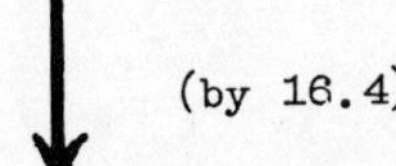

$$f_1 \sim f_2.$$

23.2   In the event the 2–manifolds X_1 and X_2 are *orientable*, the same implication diagram holds but, in addition, it is possible to replace the fourth statement by the following diagram:

(by 16.4)

There is a monotone–light factorization, lm_i, of f_i with middle space $\mathfrak{X}$, $i = 1, 2$, such that m_1 and m_2 are o–comparable and there is an isomorphism, $\eta : H^2(X_2) \approx H^2(X_1)$, guaranteeing commutativity in the following diagram:

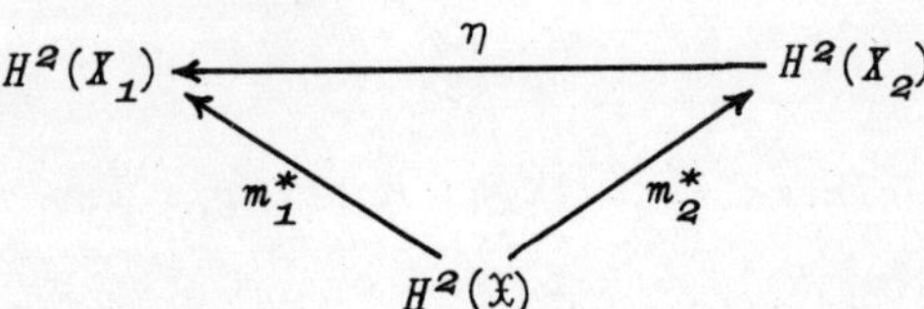

$$\downarrow \quad \text{(by 16.5)}$$

There is a monotone-light factorization, lm_i, of f_i with middle space $\mathfrak{X}$, $i = 1, 2$, such that m_1 and m_2 are o-compatible.

$$\downarrow \quad \text{(by 22.4)}$$

23.3   Given two mappings, $f_1 : H_1 \to Y$ and $f_2 : H_2 \to Y$, where H_1 and H_2 are 2-manifolds with boundary. Let X_1 and X_2 be closed 2-manifolds obtained by capping the bounding 1-spheres of H_1 and H_2 by 2-cells:

$$f_1 \sim f_2$$

$$\downarrow \quad \text{(by 15.7)}$$

There is a mapping, m_i, weakly admissible with respect to X_i, H_i, $\mathfrak{X}$ and $\mathfrak{H}$, $i = 1, 2$, such that $m_1 \sim m_2$. In addition, there is a light mapping, $l : \mathfrak{H} \to Y$, such that $f_i = l(m_i | H_i)$, $i = 1, 2$.

$$\downarrow \quad \text{(by 16.1)}$$

There is a mapping, m_i, weakly admissible with respect to X_i, H_i, $\mathfrak{X}$ and $\mathfrak{H}$, $i = 1, 2$, such that if $\mathfrak{U}$ is any normal region in $\mathfrak{X}$ while $U_i = m_i^{-1}(\mathfrak{U})$ and $\underline{m}_i = m_i | U_i$, $i = 1, 2$, then there is a homeomorphism, $h : U_1 \approx U_2$, guaranteeing commutativity in the following diagram:

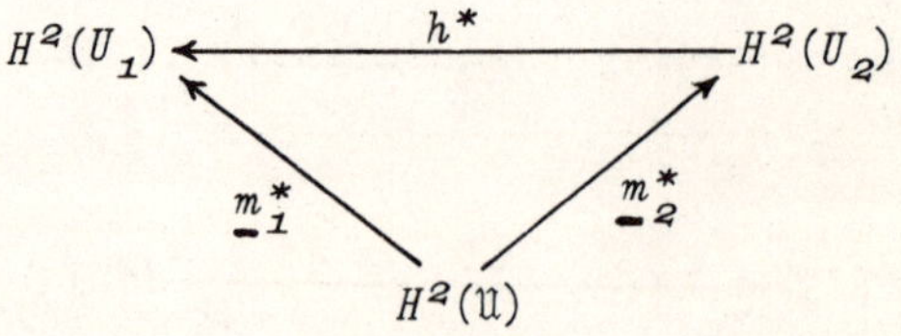

$$H^2(U_1) \xleftarrow{\quad h^* \quad} H^2(U_2)$$
$$\underline{m}_1^* \quad\nwarrow \qquad \nearrow\quad \underline{m}_2^*$$
$$H^2(\mathfrak{U})$$

In addition, there is a light mapping, $l : \mathfrak{H} \to Y$, such that $f_i = l(m_i | H_i)$, $i = 1, 2$.

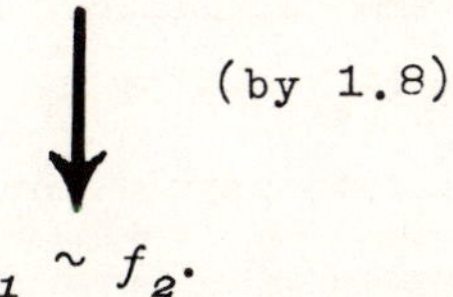

(by 16.3)

There is a mapping, m_i, weakly admissible with respect to X_i, H_i, $\mathfrak{X}$ and $\mathfrak{H}$, $i = 1, 2$, such that m_1 and m_2 are compatible. In addition, there is a light mapping, $l:\mathfrak{H} \to Y$, such that $f_i = l(m_i|H_i)$, $i = 1, 2$.

(by 22.4)

There is a mapping, m_i, weakly admissible with respect to X_i, H_i, $\mathfrak{X}$ and $\mathfrak{H}$, $i = 1, 2$, such that $(m_1|H_1) \sim (m_2|H_2)$. In addition, there is a light mapping, $l:\mathfrak{H} \to Y$, such that $f_i = l(m_i|H_i)$, $i = 1, 2$.

(by 1.8)

$$f_1 \sim f_2.$$

23.4   In the event the 2–manifolds H_1 and H_2 are *orientable*, then the same is true of X_1 and X_2. The above implication diagram holds, but in addition, it is possible to replace the fourth statement by the following diagram:

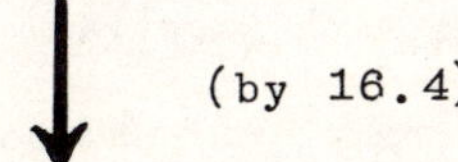

(by 16.4)

There is a mapping, m_i, weakly admissible with respect to X_i, H_i, $\mathfrak{X}$ and $\mathfrak{H}$, $i = 1, 2$, such that m_1 and m_2 are o–comparable and there is an isomorphism, $\eta:H^2(X_2) \approx H^2(X_1)$, guaranteeing commutativity in the following diagram:

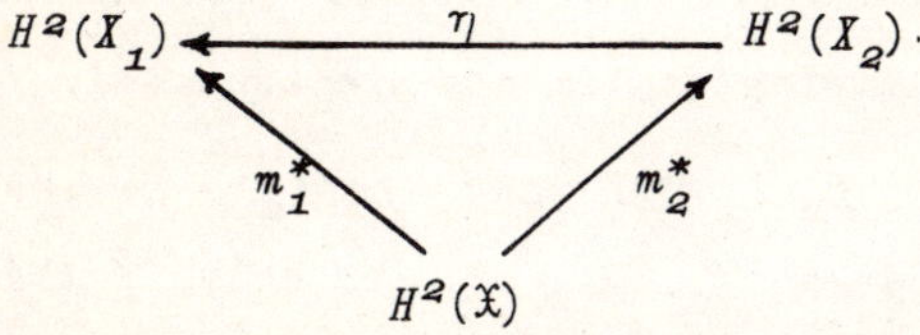

In addition, there is a light mapping, $l:\mathfrak{H} \to Y$, such that $f_i = l(m_i|H_i)$, $i = 1, 2$.

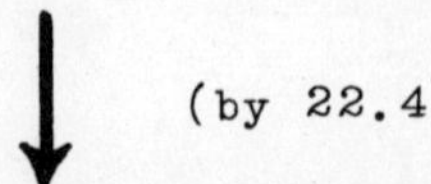

(by 16.5)

There is a mapping, m_i, weakly admissible with respect to X_i, H_i, $\mathfrak{X}$ and $\mathfrak{H}$, $i = 1, 2$, such that m_1 and m_2 are o-compatible. In addition, there is a light mapping, $l:\mathfrak{H} \to Y$, such that $f_i = l(m_i|H_i)$, $i = 1, 2$.

(by 22.4)

23.5   In connection with theorem 15.7, suppose one is given two mappings, $f_1:H_1 \to Y$ and $f_2:H_2 \to Y$, where H_1 and H_2 are 2–manifolds with boundary With the interpretation given to X_1 and X_2 in item 2° of theorem 15.7, suppose conditions 3° and 4° are fulfilled. Then m_1 and m_2 are compatible by 16.3, whence $(m_1|H_1) \sim (m_2|H_2)$ by 22.4, and $f_1 \sim f_2$ by 1.8. Consequently, one has the following *reduction* result, promised in 15.7.

THEOREM:

Given:

1°. *Two mappings, $f_1:H_1 \to Y$ and $f_2:H_2 \to Y$, where H_1 and H_2 are 2–manifolds with boundary.*

2°. *X_i is a closed 2–manifold obtained from H_i by capping the bounding Jordan curves with 2–cells, $i = 1, 2$.*

Conclusion: *$f_1 \sim f_2$ if and only if:*

3°. *There is a mapping, m_i, weakly admissible with respect to X_i, H_i, $\mathfrak{X}$ and $\mathfrak{H}$, $i = 1, 2$, such that $m_1 \sim m_2$.*

4°. *There is a light mapping, $l:\mathfrak{H} \to Y$, such that $f_i = l(m_i|H_i)$, $i = 1, 2$.*

23.6   Using the results of 23.2 together with the two theorems of 14.10, it is clear that the following statements are true (cf. Youngs [17]).

THEOREM: *If $f_1:X_1 \to Y$ and $f_2:X_2 \to Y$ are mappings, where X_1 and X_2 are 2–spheres, then $f_1 \sim f_2$ if and only if there is a monotone–light factorization, lm_i, of f_i with middle space $\mathfrak{X}$, $i = 1, 2$, and an isomorphism, $\eta:H^2(X_2) \approx H^2(X_1)$, such that for each true cyclic element $\mathfrak{E}$ of*

$\mathfrak{X}$ *commutativity holds in the following diagram where* $r:\mathfrak{X} \rightrightarrows \mathfrak{S}$ *is the monotone retraction:*

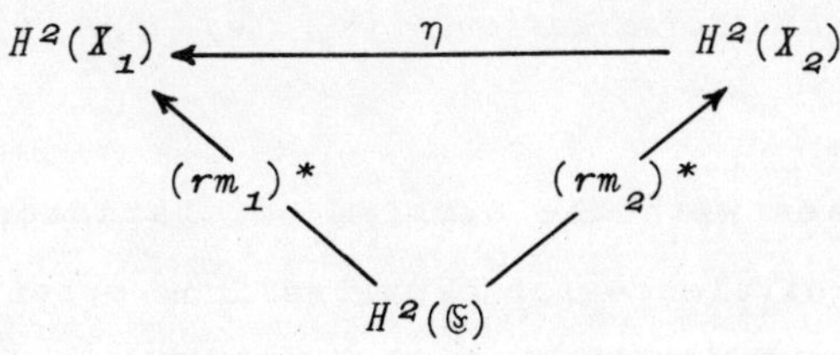

THEOREM: *If* $f_1:X_1 \to Y$ *and* $f_2:X_2 \to Y$ *are mappings, where* X_1 *and* X_2 *are 2-spheres, then* $f_1 \sim f_2$ *if and only if there is a monotone–light factorization,* lm_i, *of* f_i *with middle space* $\mathfrak{X}$, $i = 1, 2,$ *and either* m_1 *and* m_2 *are o–consistent, or there is an isomorphism,* $\eta:H^2(X_2) \approx H^2(X_1)$, *such that commutativity holds in the following diagram:*

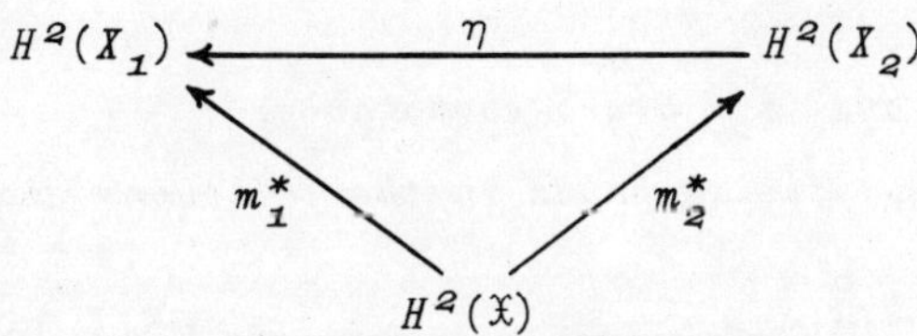

24. Oriented Fréchet equivalence

A useful concept in analysis is a modification of Fréchet equivalence due to McShane [7] called oriented Fréchet equivalence. A few comments will be made on this topic.

24.1   In the interests of generality, suppose $f_1:X_1 \to Y$ and $f_2:X_2 \to Y$ are mappings from Peano spaces X_1 and X_2. In addition, suppose that one is given an isomorphism, $\eta:H^q(X_2) \approx H^q(X_1)$, for some integer $q \geq 0$. The notation $f_1 \overset{\eta}{\sim} f_2$ (read f_1 is oriented Fréchet equivalent to f_2 with regard to η) means that there is a sequence, $\{h_n\}$, of homeomorphisms, $h_n:X_1 \approx X_2$, $n = 1, 2, 3, \cdots$, such that $f_2 h_n \rightrightarrows f_1$ and the induced isomorphism, $h_n^*:H^q(X_2) \approx H^q(X_1)$, is η for $n = 1, 2, 3, \cdots$.

24.2   A variation of the reduction theorem of 1.8 is required for the discussion to follow. (The proof is an obvious modification of the proof in Youngs [11].)

THEOREM: *If $f_1:X_1 \to Y$ and $f_2:X_2 \to Y$ are mappings from Peano spaces, and $\eta:H^q(X_2) \approx H^q(X_1)$ for some $q \geq 0$, then $f_1 \overset{\eta}{\sim} f_2$ if and only if there is a monotone–light factorization, lm_i, of f_i, $i = 1, 2$, such that $m_1 \overset{\eta}{\sim} m_2$.*

24.3   In connection with the problem of characterizing the concept of oriented Fréchet equivalence for mappings from orientable 2–manifolds, for reasons of brevity, a single criterion will be given. In view of the other results tabulated in this chapter, there is no difficulty in obtaining several characterizations.

THEOREM: *If $f_1:X_1 \to Y$ and $f_2:X_2 \to Y$ are mappings from closed orientable 2–manifolds and $\eta:H^2(X_2) \approx H^2(X_1)$, then $f_1 \overset{\eta}{\sim} f_2$ if and only if there is a monotone–light factorization, lm_i, of f_i with middle space $\mathfrak{X}$, $i = 1, 2$, such that:*

    1° *m_1 and m_2 are o–comparable.*

    2° *Commutativity holds in the following diagram:*

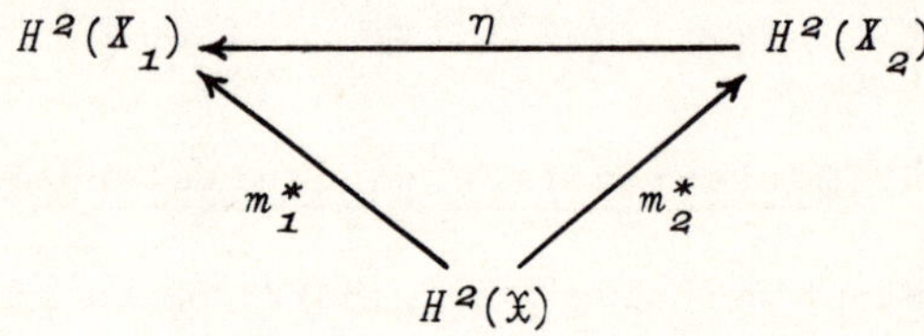

Proof: If $f_1 \overset{\eta}{\sim} f_2$, then by 24.1 there is a monotone–light factorization, lm_i, of f_i with middle space $\mathfrak{X}$, $i = 1, 2$, such that $m_1 \overset{\eta}{\sim} m_2$. In view of 16.6, this means that m_1 and m_2 satisfy conditions 1° and 2° above.

Conversely, suppose lm_i is a monotone–light factorization of f_i with middle space $\mathfrak{X}$, $i = 1, 2$, such that m_1 and m_2 satisfy conditions 1° and 2°. In view of the proof of 16.5, this implies that m_1 and m_2 are o–compatible while, relative to o–consistency, η is a matching isomorphism. Hence $m_1 \overset{\eta}{\sim} m_2$ by 22.5. Therefore $f_1 \overset{\eta}{\sim} f_2$ by 24.1.

25. Concluding remarks

The solution of the representation problem offered in these pages should serve to clarify the rather pessimistic comments made in 2.1 as to the possibility of finding a solution in the event, for example, the given mappings are from 3-cells.   A comparable line of attack would call for a characterization of the possible monotone images of a 3-cell, and this would appear to depend upon a solution of the outstanding topological problem of characterizing a 3-cell.

With the results in hand for 2-manifolds, however, there appears to be every possibility of extending to Fréchet surfaces of the type of a 2-manifold (see 2.1) some of the results on Lebesgue area which have been proved heretofor only for Fréchet surfaces of the type of a 2-cell or 2-sphere.   Thus this deeper insight into the "structure" of a Fréchet surface should also pay dividends in analysis.

Bloomington, Indiana

January 23, 1951

BIBLIOGRAPHY

1. K. Borsuk, *Sur l'addition homologique des types de transformations continues en surfaces sphériques*, Annals of Mathematics vol. 38 (1937) pp. 733–738.

2. H. Cartan, *Méthodes modernes en topologie algébrique*, Commentarii Mathematici Helvitici vol. 18 (1945) pp. 1–15.

3. S. Eilenberg and N. E. Steenrod, *Axiomatic approach to homology theory*, Proceedings of the National Academy of Sciences, U.S.A. vol. 31 (1945) pp. 117–120.

4. W. Hurewicz and H. Wallman, *Dimension theory*, Princeton University Press (1941).

5. B. v. Kerékjártó, *Involutions et surfaces continues*, Acta Universitatis Szegediensis vol. 3 (1927) pp. 46–67.

6. S. Lefschetz, *Algebraic topology*, American Mathematical Society Colloquium Publications vol. 27 (1942).

7. E. J. McShane, *On the semi-continuity of double integrals in the calculus of variations*, Annals of Mathematics vol. 33 (1942) pp. 460–484.

8. T. Radó, *On continuous mappings of Peano spaces*, Transactions of the American Mathematical Society vol. 58 (1945) pp. 420–454.

9. J. H. Roberts and N. E. Steenrod, *Monotone transformations of two-dimensional manifolds*, Annals of Mathematics vol. 39 (1938) pp. 851–862.

10. G. T. Whyburn, *Analytic topology*, American Mathematical Society Colloquium Publications vol. 28 (1942).

11. J. W. T. Youngs, *A reduction theorem concerning the representation problem for Fréchet varieties*, Proceedings of the National Academy of Sciences, U.S.A. vol. 32 (1946) pp. 328–330.

12. ———————, *Homeomorphic approximations to monotone mappings,* Duke Mathematical Journal vol. 15 (1948) pp. 87–94.

13. ———————, *K-cyclic elements,* American Journal of Mathematics vol. 62 (1940) pp. 449–456.

14. ———————, *Lebesque, Fréchet and Kerékjártó varieties,* American Journal of Mathematics vol. 70 (1948) pp. 481–496.

15. ———————, *Remarks on cyclic additivity,* Bulletin of the American Mathematical Society vol. 55 (1949) pp. 427–432.

16. ———————, *The extension of a homeomorphism defined on the boundary of a 2-manifold,* Bulletin of the American Mathematical Society vol. 54 (1948) pp. 805–808.

17. ———————, *The topological theory of Fréchet surfaces,* Annals of Mathematics vol. 45 (1944) pp. 753–785.

INDEX OF TERMS

The numbers following the terms indicate the paragraphs in and pages on which the concepts are defined. The reader is asked to consult Whyburn [10] for cyclic element theory, Cartan [2] for algebraic topology, and Youngs [12] for the notation and terminology employed wherever this paper is concerned with upper semi—continuous collections.

INDEX OF SYMBOLS